ZIML Math Competition Book

Junior Varsity 2016-2017

Areteem Institute

Edited by   John Lensmire
David Reynoso
Kevin Wang
Kelly Ren

Cover and chapter title photographs by Kelly Ren and Kevin Wang

PUBLISHED BY ARETEEM PUBLISHING
WWW.ARETEEM.ORG

ISBN: 1-944863-13-3
ISBN-13: 978-1-944863-13-5

First printing, March 2018.

Contents

Introduction

Each month during the school year, Areteem Institute hosts the online Zoom International Math League (ZIML) competitions. Students can compete in one of five divisions based on their age and mathematical level.

The book contains the problems, answers, and full solutions from the nine ZIML Junior Varsity Competitions held during the 2016-2017 School Year. It is divided into three parts:

1. The complete Junior Varsity ZIML Competitions (20 questions per competition) from October 2016 to June 2017.
2. The solutions for each of the competitions, including detailed work and helpful tricks.
3. An appendix including the topics and knowledge points covered for Junior Varsity, a glossary including common mathematical terms, and answer keys for each of the competitions so students can easily check their work.

The questions found on the ZIML competitions are meant to test your problem solving skills and train you to apply the knowledge you know to many different applications. We hope you enjoy the problems!

About Zoom International Math League

The Zoom International Math League (ZIML) has a simple goal: provide a platform for students to build and share their passion for math and other STEM fields with students from around the globe. Started in 2008 as the Southern California Mathematical Olympiad, ZIML has a rich history of past participants who have advanced to top tier colleges and prestigious math competitions, including American Math Competitions, MATHCOUNTS, and the International Math Olympaid.

The ZIML Core Online Programs, most available with a free account at ziml.areteem.org, include:

- **Daily Magic Spells:** Provides a problem a day (Monday through Friday) for students to practice, with full solutions available the next day.
- **Weekly Brain Potions:** Provides one problem per week posted in the online discussion forum at ziml.areteem.org. Usually the problem does not have a simple answer, and students can join the discussion to share their thoughts regarding the scenarios described in the problem, explore the math concepts behind the problem, give solutions, and also ask further questions.
- **Monthly Contests:** The ZIML Monthly Contests are held the first weekend of each month during the school year (October through June). Students can compete in one of 5 divisions to test their knowledge and determine their strengths and weaknesses, with winners announced after the competition.
- **Math Competition Practice:** The Practice page contains sample ZIML contests and an archive of AMC-series tests for online practice. The practices simulate the real contest environment with time-limits of the contests automatically controlled by the server.
- **Online Discussion Forum:** The Online Discussion Forum

is open for any comments and questions. Other discussions, such as hard Daily Magic Spells or the Weekly Brain Potions are also posted here.

These programs encourage students to participate consistently, so they can track their progress and improvement each year.

In addition to the online programs, ZIML also hosts onsite Local Tournaments and Workshops in various locations in the United States. Each summer, there are onsite ZIML Competitions at held at Areteem Summer Programs, including the National ZIML Convention, which is a two day convention with one day of workshops and one day of competition.

ZIML Monthly Contests are organized into five divisions ranging from upper elementary school to advanced material based on high school math.

- **Varsity:** This is the top division. It covers material on the level of the last 10 questions on the AMC 12 and AIME level. This division is open to all age levels.
- **Junior Varsity:** This is the second highest competition division. It covers material at the AMC 10/12 level and State and National MathCounts level. This division is open to all age levels.
- **Division H:** This division focuses on material from a standard high school curriculum. It covers topics up to and including pre-calculus. This division will serve as excellent practice for students preparing for the math portions of the SAT or ACT. This division is open to all age levels.
- **Division M:** This division focuses on problem solving using math concepts from a standard middle school math curriculum. It covers material at the level of AMC 8 and School or Chapter MathCounts. This division is open to all students who have not started grade 9.

- **Division E:** This division focuses on advanced problem solving with mathematical concepts from upper elementary school. It covers material at a level comparable to MOEMS Division E. This division is open to all students who have not started grade 6.

This problem book features the Junior Varsity Contests. For a detailed list of topics covered for Junior Varsity see p.135 in the Appendix.

About Areteem Institute

Areteem Institute is an educational institution that develops and provides in-depth and advanced math and science programs for K-12 (Elementary School, Middle School, and High School) students and teachers. Areteem programs are accredited supplementary programs by the Western Association of Schools and Colleges (WASC). Students may attend the Areteem Institute through these options:

- Live and real-time face-to-face online classes with audio, video, interactive online whiteboard, and text chatting capabilities;
- Self-paced classes by watching the recordings of the live classes;
- Short video courses for trending math, science, technology, engineering, English, and social studies topics;
- Summer Intensive Camps on prestigious university campuses and Winter Boot Camps;
- Practice with selected daily problems for free, and monthly ZIML competitions at ziml.areteem.org.

The Areteem courses are designed and developed by educational experts and industry professionals to bring real world applications into STEM education. The programs are ideal for students who wish to build their mathematical strength in order to excel academically and eventually win in Math Competitions (AMC, AIME, USAMO, IMO, ARML, MathCounts, Math Olympiad, ZIML, and other math leagues and tournaments, etc.), Science Fairs (County Science Fairs, State Science Fairs, national programs like Intel Science and Engineering Fair, etc.) and Science Olympiad, or purely want to enrich their academic lives by taking more challenges and developing outstanding analytical, logical thinking and creative problem solving skills.

Since 2004 Areteem Institute has been teaching with methodology that is highly promoted by the new Common Core State Standards: stressing the conceptual level understanding of the math concepts, problem solving techniques, and solving problems with real world applications. With the guidance from experienced and passionate professors, students are motivated to explore concepts deeper by identifying an interesting problem, researching it, analyzing it, and using a critical thinking approach to come up with multiple solutions.

Thousands of math students who have been trained at Areteem achieved top honors and earned top awards in major national and international math competitions, including Gold Medalists in the International Math Olympiad (IMO), top winners and qualifiers at the USA Math Olympiad (USAMO/JMO), and AIME, top winners at the Zoom International Math League (ZIML), and top winners at the MathCounts National. Many Areteem Alumni have graduated from high school and gone on to enter their dream colleges such as MIT, Cal Tech, Harvard, Stanford, Yale, Princeton, U Penn, Harvey Mudd College, UC Berkeley, UCLA, etc. Those who have graduated from colleges are now playing important roles in their fields of endeavor.

Further information about Areteem Institute, as well as updates and errata of this book, can be found online at http://www. areteem.org.

Acknowledgments

This book contains the Online ZIML Junior Varsity Problems from the 2016-17 school year. These problems were created and compiled by the staff of Areteem Institute. These problems were inspired by questions from the Areteem Math Challenge Courses, past questions on the ACT/SAT/GRE, past math competitions, math textbooks, and countless other resources and people encountered by the Areteem Curriculum Department in their life devoted to math. We thank all these sources for growing and nurturing our passion for math.

The Areteem staff, including John Lensmire, David Reynoso, Kevin Wang, and Kelly Ren, are the main contributors who compiled, edited, and reviewed this book. Photographs included on the cover and chapter introduction pages are credit to Kelly Ren and Kevin Wang.

Lastly, thanks to all the students who have participated and continue to participate in the Zoom International Math League. Your dedication to the Daily Magic Spells and Monthly Contests makes all of this possible, and we hope you continue to enjoy ZIML for years to come!

1. ZIML Contests

This part of the book contains the Junior Varsity ZIML Contests from the 2016-17 School Year. There were nine monthly competitions, held on the dates found below:

- October 7-8
- November 4-6
- December 2-4
- January 6-8
- February 3-5
- March 3-5
- April 7-9
- May 5-7
- June 2-4

1.1 ZIML October 2016 Junior Varsity

Below are the 20 Problems from the Junior Varsity ZIML Competition held in October 2016.
The answer key is available on p.148 in the Appendix.
Full solutions to these questions are available starting on p.68.

Problem 1
A five-digit number has five distinct digits, and it is divisible by 9. What is the largest such number?

Problem 2
Suppose you have a group of 8 people. How many different photographs are there of everyone lined up if 3 of the people are identical triplets who have dressed identically?

Problem 3
Solve the equation $\sqrt{|x|+x} = 8$. What is the sum of all real solutions?

Problem 4
Suppose a sequence has recursive definition $a_1 = 2$ and $a_{n+1} = 4^{a_n} \pmod{11}$. Find a_{100}.

Problem 5
In triangle ABC, $AC = 45$, $BC = 50$, and $AB = 85$. What is the altitude on $\overline{BC}$?

Problem 6

Let x_1, x_2 be the two roots for equation $x^2 + x - 3 = 0$, find the value of $x_1^3 - 4x_2^2 + 16$.

Problem 7

Given that one of the angles of the triangle with sides 5, 7, and 8 is $60°$, find the largest angle (measured in degrees) of a triangle with sides 3, 5, and 7.

Problem 8

A two digit number equals three times the sum of its digits. What is this two digit number?

Problem 9

Suppose $\Omega = \{1, 2, 3, 4\}$, and the probability of each number in Ω is half the probability of the previous value (so, for example $P(3) = P(2)/2$). The probability of getting an odd number is M/N for integers M, N with $\gcd(M, N) = 1$. What is $M + N$?

Problem 10

How many solutions are there to $\dfrac{1}{x} - \dfrac{1}{y} = \dfrac{1}{3}$ if x and y are allowed to be any integers (positive or negative)?

Problem 11

The shortest distance from the point $P = (8, -5)$ to the line $3x + 4y + 6 = 0$ is $\sqrt{D}$ for an integer D. What is D?

Problem 12
How many ways are there to put 6 identical balls in 3 numbered boxes, so that each box gets at least one ball?

Problem 13
Consider the equation $(2x^2 - 3x + 1)^2 = 22x^2 - 33x + 1$. Find the difference between the largest and smallest solution. Express your answer as a decimal.

Problem 14
Suppose $\triangle ABC$ is a non-degenerate triangle with integer side lengths and let $\overline{AD}$ be an angle bisector. If $BD = 4$ and $DC = 5$, find the smallest possible perimeter of $\triangle ABC$.

Problem 15
The quadratic $x^2 - (2m - 3)x + m(m - 3) = 0$ has one positive and one negative root for all m such that $a < m < b$. What is $b - a$?

Problem 16
Suppose two perpendicular chords intersect and divide each other in a ratio of $1 : 2$. The radius of the circle if each chord is 12 long can be written as $A\sqrt{B}$ in simplest radical form. What is $A \times B$?

Problem 17

Consider the product of all distinct positive divisors of 60^4. This product is of the form 60^K for some integer K. What is K?

Problem 18

Suppose you have a square $ABCD$ with side length 2. Let E be a randomly chosen point inside the square. Let p denote the probability that the triangle $\triangle ABE$ is obtuse. Then $p \approx K\%$ where K is an integer multiple of 5. What is K?

Problem 19

Let a, b, c, and d be the roots of $x^4 - 2x - 1990 = 0$. Then $1/a + 1/b + 1/c + 1/d = 1/K$ for an integer K. What is K?

Problem 20

Suppose 3 different numbers are chosen from the set $\{1, 2, \ldots, 7\}$. Call two of the numbers chosen "neighbors" if they differ by exactly 1. For example, if the subset $\{1, 4, 5\}$ is chosen, then 4 and 5 are neighbors, but 1 is not neighbors with the other elements. In how many ways can this be done such that the chosen subset has at least one pair of neighbors?

1.2   ZIML November 2016 Junior Varsity

Below are the 20 Problems from the Junior Varsity ZIML Competition held in November 2016.
The answer key is available on p.149 in the Appendix.
Full solutions to these questions are available starting on p.74.

Problem 1
$a^{32} - b^{32}$ can be completely factored into binomials with integer coefficients. How many binomials (counting repeats) appear in this factorization? For example, both $(x+2)(x+1)(x-3)$ and $(x-1)^2(x^2+2)$ contain 3 binomials.

Problem 2
Suppose you write out the numbers $1 - 10000$: $1,2,3,4,\ldots,10000$. How many digits would you write in total?

Problem 3
Let $f(x) = x^4 - 13x^2 + 12$. For how many integers n is $f(n) < 0$?

Problem 4
Consider two-digit numbers with the following property: when the number is squared the last two digits are the original number. What is the sum of all possible numbers with this property?

Problem 5

Consider a solid unit cube, with 8 vertices. Consider paths that start and end at the same vertex, visit each vertex exactly once, and travel in a straight line on the outside of the cube between each vertex. The length of the longest such path is $A + B\sqrt{2}$, where A and B are integers (set $A = B = 0$ if no such path exists). What is B?

Problem 6

Given that a and b are both prime numbers and $p = a^b + b^a$ is also prime. What is p?

Problem 7

Suppose you have line segments of length $1, a, b$, where a, b are real numbers chosen randomly between 0 and 2. What is the probability you can form a triangle using the three line segments? Express your answer as $P\%$ where P is decimal. If necessary, round to the nearest hundredth. For example, the probability $\frac{1}{3}$ would be entered as 33.33.

Problem 8

Use the digits $1, 2, 3, 4, 5, 6, 7, 8, 9$ (each once) to form the largest possible number that is a multiple of 11. What is this number?

Problem 9

Suppose you have a circle with diameter $\overline{AB}$ with $AB = 12$. Let C, D be on arc $\widehat{AB}$ such that $\widehat{AC} : \widehat{CD} : \widehat{DB} = 1 : 4 : 1$. The area of the figure enclosed by line segment $\overline{AC}$, arc $\widehat{CD}$, and line segment $\overline{AD}$ can be written as $K \times \pi$ for an integer K. What is K?

Problem 10

Consider integers m such that $x^2 + (3m - 2)x - m(3 - m) = 0$ has one positive and one negative root. What is the sum of all such m?

Problem 11

Find the remainder when 4^{2016} is divided by 11.

Problem 12

Distinct real numbers a and b satisfy $(a+1)^2 = 3 - 3(a+1)$ and $3(b+1) = 3 - (b+1)^2$. Find the value of $b\sqrt{\dfrac{b}{a}} + a\sqrt{\dfrac{a}{b}}$.

Problem 13

$ABCD$ is an isosceles trapezoid (so it is cyclic) inscribed in a circle of radius 10. If the center of the circle is in the interior of the trapezoid, $AD = 12$, and $BC = 16$, what is the area of trapezoid $ABCD$?

Problem 14

How many trailing 0's are there when 1000! is multiplied out?

Problem 15

Suppose you roll a fair six-sided die three times. The probability the sum of the three rolls is 9 can be written as $\dfrac{P}{Q}$ where $\gcd(P, Q) = 1$. What is $P + Q$?

Problem 16

Let quadrilateral $ABCD$ be as in the diagram below, where $\angle BAE = \angle CAE$ and $\angle DAF = \angle CAF$.

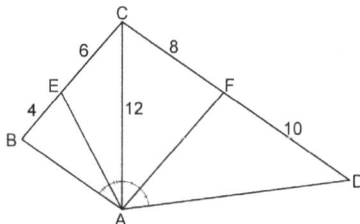

Find the perimeter of $ABCD$.

Problem 17

Let $x, y = 3 \pm \sqrt{5}$. Then

$$\frac{x^2 + xy + y^2}{x^3 y^2 + x^2 y^3 + x + y} = \frac{N}{M}$$

where $\gcd(N, M) = 1$. What is $N + M$?

Problem 18

How many ways are there to arrange the numbers $15, 25, 35, 45, 55$ so that the sum of each consecutive group of 3 numbers is divisible by 3?

Problem 19

A 30-60-90 triangle is circumscribed about a circle of radius 1. A circle is then circumscribed about this triangle. This larger circle has area $(A + \sqrt{B})$ times as large as the original circle (where A and B are integers). What is $A + B$?

Problem 20

Suppose 4 friends go to a party. They each wear a coat. However, as they are leaving, they each randomly grab a coat. How many ways can the friends leave so that NONE of them have their own coat?

1.3   ZIML December 2016 Junior Varsity

Below are the 20 Problems from the Junior Varsity ZIML Competition held in December 2016.
The answer key is available on p.150 in the Appendix.
Full solutions to these questions are available starting on p.82.

Problem 1
Suppose a man has 4 sons and 3 daughters, and has 3 boys' schools and 4 girls' schools available to choose from. How many different ways can he send his children to school so none of his daughters attend the same school?

Problem 2
Consider the equation

$$\frac{|x+4|}{|x+1|} = \frac{|x+3|}{|x+2|}.$$

The sum of all RATIONAL solutions to this equation can be written as $\dfrac{P}{Q}$ where $\gcd(P,Q) = 1$. What is $P+Q$?

Problem 3

Let $ABCD$ be a parallelogram as in the diagram, with E the midpoint of $\overline{BC}$.

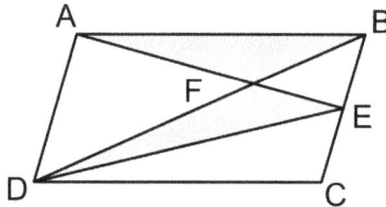

Find the combined area of the shaded regions $\triangle ABF$ and $\triangle DEF$ if the entire parallelogram has area 30.

Problem 4

Consider five-digit numbers $\overline{x123y}$ that are divisible by 36. What is the difference between the largest and smallest such number?

Problem 5

Let $\triangle ABC$ have perimeter 35. Let AE be the angle bisector of $\angle A$, with $BE = 4$ and $CE = 10$. Find the length of AC.

Problem 6

Find the remainder when $3^3 + 3^{3^2} + 3^{3^3} + \cdots + 3^{3^8}$ is divided by 11.

Problem 7

The sum of all real solutions to $4\sqrt{x^2 - 2x + 5} - 4\sqrt{x^2 - 2x + 4} = 1$ is the integer K. What is K?

Problem 8

Suppose you roll a pair of fair dice. The probability that one of the dice comes up 2, given that the sum of the numbers on the dice is 6, is $K\%$, where K is rounded to the nearest integer (if necessary). What is K?

Problem 9

Suppose you have a circle $(x - 1)^2 + (y - 1)^2 = 4$ and a line $x + y = 4$. Let C be the center of the circle and suppose the line intersects the circle at points A and B. Find the area of $\triangle ABC$.

Problem 10

Let ω_1 and ω_2 be circles defined by

$$(x + 6)^2 + (y - 8)^2 = 16$$

and

$$(x - 9)^2 + (y + 12)^2 = 36,$$

respectively. The length of the longest line segment $\overline{PQ}$ that is tangent to ω_1 at P and to ω_2 at Q is $\sqrt{M}$ for an integer M. What is M?

Problem 11

There is one ordered triple (x, y, z) where x, y, z are distinct prime numbers satisfying the equation $x(x + y) = z + 120$. What is $x + y + z$?

Problem 12

Suppose Alice, Bob, and Charlie plan to meet for dinner. They each randomly show up to dinner between 5:00 and 6:00 PM. Charlie is impatient and will only stay if Alice and Bob have already arrived. Alice is very patient and will wait for Bob or Charlie (or both). Bob will wait for Charlie if Alice has already arrived, but will not wait for Alice. The probability they have dinner together can be written as $\dfrac{N}{M}$ with $\gcd(N, M) = 1$. What is $M - N$?

Problem 13

Consider numbers n with the following property: the product of the factors of n is equal to n^2. How many such n are there with $1 \leq n \leq 30$? Recall that 1 and n are always factors of n.

Problem 14

Suppose you pick 4 vertices of a cube with side length 6 to form a tetrahedron. Find the volume of the largest such tetrahedron.

Problem 15
In the polynomial

$$(17+x)(8+x^2)(3+x^4)(9+x^8)(2+x^{16})(11+x^{32}),$$

what is the coefficient of x^{53}?

Problem 16
How many ordered pairs (x,y) of real numbers satisfy both $x^2 - xy + y^2 = 13$ and $x - xy + y = -5$?

Problem 17
Suppose you have 8 red cards and 16 black cards. Assume all the cards of the same color are identical. Deal the cards out in a line. How many arrangements of the cards are there if there must be at least 2 black cards between any two of the red cards?

Problem 18
Consider all m such that the absolute value of the difference between the roots of $2x^2 - mx - 8 = 0$ is $m - 1$. What is the sum of all such m? (If no such m exist, the sum is 0.)

Problem 19
Suppose that students take three tests in a course and that exactly 10 students get A's on each of the three exams. For any collection of two of the exams, exactly 8 students got A's on both. If 15 students got an A on at least one exam, how many students got A's on all three exams?

Problem 20

Every number less than 10000 can be written as $\overline{abcd}$ where $a, b, c, d \in \{0, 1, \ldots, 9\}$. For how many of these numbers is it true that 11 divides the sum of $\overline{ab}$ and $\overline{cd}$.

1.4 ZIML January 2017 Junior Varsity

Below are the 20 Problems from the Junior Varsity ZIML Competition held in January 2017.

The answer key is available on p.151 in the Appendix.

Full solutions to these questions are available starting on p.90.

Problem 1
In a right triangle $\triangle ABC$ suppose $\angle B = 90°$ and $\angle C = 30°$. Suppose point D is on $\overline{BC}$ with $\angle ADB = 45°$ and $DC = 2$. Find the length of AB rounded to the nearest integer.

Problem 2
Consider six-digit numbers of the form $\overline{abcdef}$ with no repeated digits using only 1, 2, 3, 4, 5, or 6. Assume that $6 \mid \overline{abcdef}$, $5 \mid \overline{abcde}$, $4 \mid \overline{abcd}$, $3 \mid \overline{abc}$, and $2 \mid \overline{ab}$. Find the difference between the largest and smallest possible such numbers.

Problem 3
Two identical rooks are placed on an 8×8 chessboard so that they are in different rows and columns. How many such arrangements are there?

Problem 4

In the figure below, the middle $1 \times 1 \times 1$ cube is removed from a $3 \times 3 \times 3$ cube.

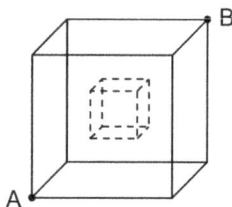

The length of the shortest path from A to B, avoiding the interior of the removed cube, is $\sqrt{T}$ for an integer T. What is T?

Problem 5

Consider solutions to the equation $\dfrac{1}{2x^2 - 3} - 8x^2 + 12 = 0$. The largest such solution can be written in the form $\sqrt{\dfrac{A}{B}}$, where A and B are relatively prime positive integers. What is $A + B$?

Problem 6

Consider lines $\ell_1 : y = x - 2$, $\ell_2 : y = x + 2$, and $\ell_3 : y = 6 - x$. Consider the largest circle such that (i) the center of the circle is on line ℓ_3, and (ii) the circle does not go outside line ℓ_1 and ℓ_2. This circle has equation $x^2 + y^2 + Ax + By + C = 0$ for integers A, B, C. What is C?

Problem 7
Find the remainder when the sum $n = 2 + 212 + 21212 + \cdots +$ 21212121212 (the last term has five 21's and a 2) is divided by 11.

Problem 8
How many rearrangements of the word *MACHINES* have no vowels next to each other?

Problem 9
How many solutions to $\dfrac{1}{a} + \dfrac{1}{b} + \dfrac{1}{c} = 1$ are there with a, b, c integers satisfying $0 < a \leq b \leq c$?

Problem 10
How many rational solutions does

$$(x^2 + 4x + 8)^2 + 3x(x^2 + 4x + 8) + 2x^2 = 0$$

have?

Problem 11
Given positive integers $1, 2, 3, \ldots, 10$. Let a permutation of these numbers satisfy the requirement that, for each number, it is either (i) greater than all the numbers after it, or (ii) less than all the numbers after it. How many such permutations are there? For example, $1, 2, 3, 4, 5, 6, 7, 8, 9, 10$ and $10, 1, 2, 3, 4, 5, 6, 7, 8, 9$ both work.

Problem 12

Let $a \neq b$ with $a, b \neq 1$ satisfy

$$a^3 + 4a^2 - 9a + 4 = 0$$

and

$$b^3 + 4b^2 - 9b + 4 = 0.$$

Then $\dfrac{1}{a} + \dfrac{1}{b} = \dfrac{P}{Q}$, with $\gcd(P, Q) = 1$. What is $Q - P$?

Problem 13

For how many positive integers $n \leq 100$ is $2^n - 1$ divisible by 31?

Problem 14

There is a single pair (a, b) such that the equation

$$x^2 + 2(1 + a)x + (3a^2 + 4ab + 4b^2 + 2) = 0$$

has real roots. For this pair, what is $a + b$? If necessary, express your answer as a decimal rounded to the nearest hundredth.

Problem 15

A three-digit positive number is randomly selected. The probability that its digits are in an arithmetic progression is $P\%$. Find the value of P, rounded to the nearest integer if necessary.

Problem 16

Let $\overline{PQ}$ be a chord with $PQ = 12$. Extend $\overline{PQ}$ to a point R such that $QR = 4$. Let T be such that $\overline{RT}$ is tangent to the circle. What is RT?

Problem 17

From the set $\{1, 2, \ldots, 100\}$, select a subset of k numbers. What is the minimum value of k such that the subset is guaranteed to have two numbers with 3 as a common divisor?

Problem 18

A manufacturer has three production lines that produce a certain kind of IC chip with a failure rate of 1%. Suppose that one of the production lines malfunctions and starts producing IC chips with a failure rate of 10%. Suppose further that chips are randomly packaged and shipped out before the malfunction is noticed. If you get a chip from this lot, the probability it is defective is $\dfrac{N}{M}$ where $\gcd(N, M) = 1$. What is $N + M$?

Problem 19

Let G be the centroid of $\triangle ABC$, with $AG = 5$, $BG = 12$, and $CG = 13$. Find $[ABC]$ (where $[ABC]$ means the area of triangle ABC).

Problem 20

For how many integers k with $|k| \leq 5$ does $\sqrt{2x^2 + 4} = x + k$ have real solutions?

1.5 ZIML February 2017 Junior Varsity

Below are the 20 Problems from the Junior Varsity ZIML Competition held in February 2017.

The answer key is available on p.152 in the Appendix.

Full solutions to these questions are available starting on p.98.

Problem 1

$x^7 + x^6 + x^5 + x^4 + x^3 + x^2 + x + 1$ can be completely factored into binomials. How many binomials (counting repeats) appear in this factorization? For example, both $(x^2 + 2)(x^2 + 1)(x - 3)$ and $(x - 1)^2(x^2 + 4)$ contain 3 binomials.

Problem 2

What is the radius of the largest circle that can fit in a $60°$ sector of a circle with radius 12?

Problem 3

What is the units digit of $1^2 + 2^2 + 3^2 + \cdots + 99^2$?

Problem 4

A shop has 8 types of cookies and you want to buy 6 cookies. In how many ways can you buy the cookies if you want to make sure to have an even number of each type of cookies you buy?

Problem 5

Let x_1 and x_2 be the two roots for equation $x^2 + x - 4 = 0$. Find the value of $x_1^3 - 5x_2^2 + 25$.

Problem 6
A 5-digit number, all of whose digits are distinct, is divisible by 11. Given that its left-most digit is 3. What is the smallest such number?

Problem 7
Find the smallest perfect square that is a multiple of both 108 and 264.

Problem 8
Suppose you have the numbers 1, 2, 3, 4, 5. How many different 5-digit numbers can be formed using each digit once and with 2 next to 1 or 3?

Problem 9
Suppose $\triangle ABC$ is a right triangle with $\angle C = 90°$ and $AC = 5$, $BC = 12$. Let $\overline{AD}$ be an angle bisector. Find the area of $\triangle ABD$ rounded to the nearest integer.

Problem 10
Suppose you toss a fair (two-sided) coin. You then roll a fair die once if you get heads, and twice if you get tails. The probability your coin came up tails if the sum of your rolls is 6 is $\frac{N}{M}$, where $\gcd(N,M) = 1$. What is $M+N$?

Problem 11
Find the remainder when 26! is divided by 29.

Problem 12
There are two solutions to $|x^2 + 6x + 1| = |(x+3)^2 - 4|$, which can be written in the form $A \pm \sqrt{B}$. What is $A \cdot B$?

Problem 13
Solve the equation $\sqrt{|x| + 2x} = 12$. What is the sum of all real solutions?

Problem 14
Suppose you have a student group with 5 males and 10 females. How many ways are there to pick an Executive Committee of 4 members and a Party Planning Committee of 4 members? Members can be in both committees at once, but each committee must have at least one male and at least one female.

Problem 15
Four non-overlapping regular plane polygons all have sides of length 1. The polygons meet at a point A in such a way that the sum of the four interior angles at A is $360°$. Among the four polygons, two are squares and one is a triangle. What is the perimeter of the entire shape?

Problem 16

Suppose 12 people will divide and form 4 teams of 3 people to play two games of 3 on 3 basketball, a first game and a second game. If we only care about the two games being played (i.e. 'Team A vs Team B and Team C vs Team D' is the same as 'Team B vs Team A and Team D vs Team C') how many different ways can the games happen?

Problem 17

Find the remainder when $222^{555} + 555^{222}$ is divided by 7.

Problem 18

The altitudes of a triangle are in ratio $2 : 2 : 3$ and the triangle has a perimeter of 24. Then the area of the triangle can be written as $\sqrt{M}$ for an integer M. What is M?

Problem 19

The equation in x, $x^2 + px + q = 0$, has two nonzero integer roots. If $p + q = 198$, what is p?

Problem 20

An obtuse triangle with dimensions 9, 10, and 17 is rotated about the smallest side so that it creates a three-dimensional solid shown below.

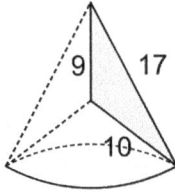

The surface area of the resulting solid is $K \cdot \pi$ for an integer K. What is K?

1.6 ZIML March 2017 Junior Varsity

Below are the 20 Problems from the Junior Varsity ZIML Competition held in March 2017.
The answer key is available on p.153 in the Appendix.
Full solutions to these questions are available starting on p.105.

Problem 1
Find the maximum value of the function $f(x) = 61 + 72x - 36x^2$.

Problem 2
If four circles are drawn in a plane, what is the maximum number of points that belong to at least two of the circles?

Problem 3
How many factors of 2^{20} are larger than $5,000$?

Problem 4
Find the last two digits of 21^{2017}.

Problem 5
The sum of squares of the roots of equation $x^2 + 2kx = 3$ is 10. If $k > 0$, find k.

Problem 6

How many solutions to the equation $a + b + c = 10$ are there with $a, b, c \geq 2$?

Problem 7

Let ω_1 and ω_2 be circles defined by

$$x^2 + (y + 10)^2 = 36$$

and

$$x^2 + (y - 15)^2 = 16,$$

respectively. Let $\sqrt{M}$ be the length of the shortest line segment $\overline{PQ}$ that is tangent to ω_1 at P and to ω_2 at Q, find M.

Problem 8

Find the number of factors of 3240.

Problem 9

The equation $x^2 + (a - 6)x + a = 0$ has two integer roots. If $a > 0$, find the value of a.

Problem 10

Find the smallest prime p such that $2^p - 1$ is not a prime number.

Problem 11
Suppose you pick 4 vertices of a cube to form a tetrahedron. How many different (non-congruent) tetrahedra are possible?

Problem 12
Suppose there are 5 red marbles, 3 blue marbles, and 3 green marbles. How many ways can you arrange the marbles in a line such that no two marbles of the same color are located next to each other and no blue marbles are located next to green marbles?

Problem 13
The two solutions of $|x^2 + 1| = 2|x - 1|$ can be expressed in the form $A \pm \sqrt{B}$, where A and B are integers. Find $A + B$.

Problem 14
Suppose we have an 8×8 checkerboard. How many ways are there to put a red checker piece and a black checker piece such that one of the checker pieces is adjacent (horizontally, vertically or diagonally) to the other checker piece?

Problem 15
Suppose that A has 9 divisors and B has 4 divisors. Find $A + B$ if $\gcd(A, B) = 7$ and $\text{lcm}(A, B) = 2205$.

Problem 16
Given square $ABCD$, let P and Q be the points outside the square that make triangles CDP and BCQ equilateral. Segments AQ and BP intersect at T. Find angle ATP in degrees.

Problem 17
What is the smallest integer root of $(x^2 + x + 1)(x^2 + x + 2) - 12 = 0$?

Problem 18
Consider a trapezoid with height 8. If the diagonals of the trapezoid are perpendicular and one of them has length 10, the area of the trapezoid can be expressed as a fraction $\frac{p}{q}$ in lowest terms. Find $p + q$.

Problem 19
A positive 8-digit integer has only 2 different digits. What is the smallest such number that is a multiple of both 5 and 6?

Problem 20

Suppose that a dealer has a standard 52-card deck and offers to play a game with you. He will pay you $1 if he draws a black card and $0 if he draws a red card. Additionally, he will pay you $1 if he draws a face card and $0 otherwise. For example, if he drew a King of Hearts, you would earn $1 since it is a red ($0) face ($1) card. The probability of earning $2 can be expressed as a fraction $\frac{p}{q}$ in lowest terms. Find $p+q$. (Recall that a standard deck of cards has 4 suits: hearts and diamonds, which are red, and clubs and spades, which are black. Each of the suits has 13 cards, labeled $2,3,\ldots,10,J,Q,K,A$. J,Q,K are face cards.)

1.7 ZIML April 2017 Junior Varsity

Below are the 20 Problems from the Junior Varsity ZIML Competition held in April 2017.
The answer key is available on p.154 in the Appendix.
Full solutions to these questions are available starting on p.113.

Problem 1
Find the smallest perfect square that is a multiple of both 360 and 525.

Problem 2
Suppose $PENTA$ is a regular pentagon. Equilateral triangle TAB is drawn sharing side $\overline{AT}$ with the pentagon, so that $\angle BAP$ is as large as possible (with $\angle BAP < 180°$). What is the measure of $\angle BAP$ in degrees?

Problem 3
Suppose 5 boys and 3 girls run in a race (with no ties). If none of the girls finish right after one another (2nd and 3rd, 5th and 6th, etc.), how many possible outcomes of the race are there?

Problem 4
Consider the polynomial $p(x) = 16 + 4x - 4x^2 - x^3$. Let r be the largest root and s be the smallest root. What is $|r| + |s|$?

Problem 5
Suppose A, B, C are points on a circle such that the angular measures of arc $\widehat{AB}$ (not containing C) and arc $\widehat{CA}$ (not containing B) are in ratio $6 : 10$. Suppose further that $\angle ABC = 80°$. Find the measure of $\angle BAC$ in degrees.

Problem 6
Consider the diagram below

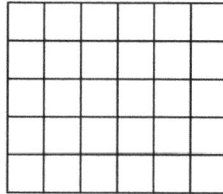

How many rectangles are in the diagram? (Remember a square is still a rectangle!)

Problem 7
Consider integer solutions to

$$(x^2 + 4x + 5)^2 + 5x(x^2 + 4x + 5) + 6x^2 = 0.$$

What is the product of all the integer solutions?

Problem 8

Suppose you have a trapezoid $ABCD$ with $\overline{AB} \parallel \overline{CD}$. Let E be the intersection of the diagonals. Suppose $AB = 10$ and $CD = 15$. Then the ratio of the area of $\triangle ADE$ to the area of the trapezoid $ABCD$ can be written as a reduced fraction $\dfrac{P}{Q}$ (with P, Q positive integers). What is $P + Q$?

Problem 9

Find the remainder when $15^{999} + 33^{100}$ is divided by 16.

Problem 10

Let x be a real number such that $x^3 + 3x = 9$. Determine the value of $x^7 + 54x(x-1)$.

Problem 11

Suppose you flip a fair coin 8 times and you know that the outcomes contains 5 tails and 3 heads. The probability that the three heads all occurred in a row is $\dfrac{P}{Q}$ with $\gcd(P,Q) = 1$ and $P, Q > 0$. What is $Q - P$?

Problem 12

Consider the number

$$201720172017\cdots 2017,$$

which consists of 2017 copies of the digits 2017. What is the remainder when this number is divided by 11?

Problem 13

Let m be an integer. The equation $x^2 + mx - m + 1 = 0$ has two distinct positive integer roots. Find m.

Problem 14

Alice, Bob, and Charlie decide to meet up at Alice's house after school. Bob randomly arrives sometime between 5 and 6 pm, while Charlie randomly arrives sometime between 5 and 7 pm. What is the probability that Charlie arrives before Bob? Express your answer as a decimal, rounded to the nearest hundredth if necessary. Note: Bob and Charlie do not necessarily arrive on a whole number minute, second, etc, so for example Charlie could arrive $\sqrt{2}$ hours after 5 pm.

Problem 15

$\triangle ABC$ is a right triangle with integer side lengths. If one of the side lengths is 17, what is the largest possible area of the triangle?

Problem 16

Let quadrilateral $ABCD$ be in the diagram below, where all the 4 marked angles are equal.

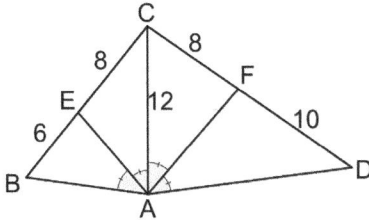

Find the perimeter of $ABCD$.

Problem 17

How many different real roots does $2x^4 - 9x^3 + 14x^2 - 9x + 2 = 0$ have?

Problem 18

The Fibonacci sequence starts with $1, 1, 2, 3, 5$, where each term is the sum of the previous two terms. For misbehaving in math class, Alex's teacher made him choose a digit $D \neq 1$, and then write the Fibonacci sequence until the digit D was the units digit of two consecutive numbers in the sequence. What digit D should Alex choose to minimize the number of elements he needs to calculate from the sequence?

Problem 19

Let $S - ABC$ be a regular tetrahedron with volume 100. Let E, F, G be the midpoints of $\overline{AB}, \overline{BC}, \overline{AC}$ and J, K, L be the midpoints of $\overline{SA}, \overline{SB}, \overline{SC}$. Form solid $EFG - JKL$. What is the volume of this solid?

Problem 20

Suppose you have one copy each of the 7 different Harry Potter books. You give out all 7 books to 3 of your friends. Each friend gets at least one book. How many ways can you give out the books? (The books you give each friend are not in any specific order.)

1.8   ZIML May 2017 Junior Varsity

Below are the 20 Problems from the Junior Varsity ZIML Competition held in May 2017.
The answer key is available on p.155 in the Appendix.
Full solutions to these questions are available starting on p.120.

Problem 1
Let $\overline{a357b}$ be a five-digit number. If the number $\overline{a357b}$ is divisible by 44, find the the five-digit number.

Problem 2
Let triangle ABC be an equilateral triangle with side length 16. Let D be on side $\overline{AB}$ and E be on side $\overline{AC}$ such that $\overline{DE} \| \overline{BC}$. Assume triangle ADE and trapezoid $DECB$ have the same perimeter. Then the area of $\triangle ADE$ is $M\sqrt{3}$ for an integer M. What is M?

Problem 3
Suppose you place 2 different rings on 4 fingers (not your thumb) on your left hand. How many different outcomes are possible?

Problem 4
Let x_1 and x_2 be the two roots of $17x^2 - 8x - 2 = 0$. Then $x_1^2 + x_2^2$ can be written as $\dfrac{P}{Q}$ for positive integers P, and Q, with $\gcd(P, Q) = 1$. What is $P + Q$?

Problem 5

How many rearrangements of the word *MACHINES* have no vowels next to each other? These rearrangements do NOT have to be actual words. Recall that the letters A, E, I, O, U are vowels.

Problem 6

What is the remainder when $1^2 + 2^2 + \cdots + 49^2$ is divided by 5?

Problem 7

Let $\overline{AM}$ be a median of $\triangle ABC$, D be a point on $\overline{MC}$, and E be a point on $\overline{AB}$, such that $\overline{ME} \parallel \overline{AD}$. If the area of $\triangle ABC$ is 90, what is the area of $\triangle BDE$.

Problem 8

If $x - \dfrac{1}{x} = 5$, then

$$x^2 + x - 14 + \frac{1-x}{x^2}$$

is an integer L. What is L?

Problem 9

How many 5-digit numbers contain the digit 5 at least once?

Problem 10
What is the smallest number with 10 factors?

Problem 11
Four identical balls (spheres), each of radius 1 in, are glued to the ground so that their centers form the vertices of a square with side length 2 in. Suppose you rest a fifth identical ball on the four balls (so the fifth ball is a sphere externally tangent to the other spheres). How many inches does this ball rest off the ground? Round your answer to the nearest tenth if necessary.

Problem 12
Consider solutions to $\sqrt{x+3} - \sqrt{3x-2} = -1$. Find the sum of all real solutions. (If there are no solutions, input a sum of 0.)

Problem 13
Suppose 10 people get in an elevator on Floor 0. Each of the ten people leave on either Floor 1, Floor 2, Floor 3, Floor 4, or Floor 5. If we only care about how many people get off at each floor, how many ways can the people get off? That is, one possibility is 3 people get off at Floor 2 and the rest get off at Floor 4 (we do not care exactly which people get off at a floor, just the number of people that do so).

Problem 14
Let $ABCD$ be a unit square and let P and Q be on sides $\overline{AD}$ and $\overline{AB}$, respectively, such that $\triangle APQ$ has perimeter 2. Find the measure of $\angle PCQ$ in degrees.

Problem 15

Let a, b, and c be the roots of $2x^3 + 20x^2 - 75x + 50 = 0$. Find the value of $\dfrac{1}{ab} + \dfrac{1}{ac} + \dfrac{1}{bc}$. Round your answer to the nearest hundredth if necessary.

Problem 16

From the set $\{1, 2, \ldots, 100\}$, select k numbers. What is the minimum value of k such that it is guaranteed to have two numbers that are not relatively prime? Hint: How many prime numbers are there less than 100?

Problem 17

There is an urn with 5 green, 6 red, and 4 yellow balls. You pick 3 balls without replacement (that is, without putting the balls back after each pick). Suppose you are given that at least 2 of the balls are green (you are not given which specific balls are green, just at least two of them are). Then the probability all 3 balls are green can be written as $\dfrac{P}{Q}$ for positive integers P, Q with $\gcd(P, Q) = 1$. What is $Q - P$?

Problem 18

Consider all integer solutions to $\dfrac{1}{x} + \dfrac{1}{y} = \dfrac{1}{5}$ with $0 < x < y$. For the solution where $\dfrac{y}{x}$ is the largest, what is $\dfrac{y}{x}$? Round your answer to the nearest tenth if necessary.

Problem 19

Let A, B, C, D be four points, arranged in clockwise order, on circle ω. Chords AC and BD intersect at P. Given that $AB = 8$, $BP = 6$, $PA = 10$, and $PC = 5.4$, find the radius of circle ω. Round your answer to the nearest tenth if necessary.

Problem 20

The system of equations $y = mx + \sqrt{2}$ and $x^2 + y^2 = 1$ has at least one pair (x, y) of real solutions for any m in the closed interval $[a, b]$ (for a, b real numbers), and no real solutions outside of this interval. What is $b - a$? Round your answer to the nearest tenth if necessary.

1.9 ZIML June 2017 Junior Varsity

Below are the 20 Problems from the Junior Varsity ZIML Competition held in June 2017.
The answer key is available on p.156 in the Appendix.
Full solutions to these questions are available starting on p.127.

Problem 1
Suppose you write out the numbers $1 - 1000$: $1, 2, 3, 4, \ldots, 1000$. How many digits have you written in total?

Problem 2
Consider solutions to $\sqrt{5-x} + \sqrt{2+x} = \sqrt{5} + \sqrt{2}$. How many real solutions are there?

Problem 3
Find the smallest number that leaves a remainder of 1 when divided by 4, 2 when divided by 5, and 3 when divided by 6.

Problem 4
In equiangular octagon $ABCDEFGH$, $AB = CD = EF = GH = 8\sqrt{2}$ and $BC = DE = FG = HA$. Given that BC is an integer and the area of the octagon is 356, compute the length of side BC.

Problem 5
A four digit number minus the sum of its digits gives the four digit number $\overline{20d0}$. What is d?

Problem 6

A six-digit number $\overline{xy342z}$ is divisible by 396. Find the smallest possible such number.

Problem 7

Given right triangle ABC, construct semicircles on the three sides as shown.

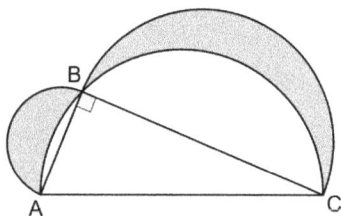

Given that $AB = 5$ and $BC = 12$, find the sum of the areas of the shaded regions. Round your answer to the nearest tenth if necessary.

Problem 8

Find the sum of all real solutions of $|x^2 - 3x + 1| = |x^2 + x + 2|$. Round your answer to the nearest hundredth if necessary.

Problem 9

A bookshelf contains 9 different books: 2 German books, 3 Spanish books , and 4 French books. The 9 books are arranged in a line, with books of the same language next to each other. How many different arrangements are there?

Problem 10

ABC is a triangle with integer side lengths. Extend $\overline{AC}$ beyond C to point D such that $CD = 125$. Similarly, extend $\overline{CB}$ beyond B to point E such that $BE = 120$ and $\overline{BA}$ beyond A to point F such that $AF = 35$. If triangles CBD, BAE, and ACF all have the same area, what is the minimum possible area of triangle ABC?

Problem 11

The equation $(a+1)x^2 + (a-1)x - 2 = 0$ has exactly one real solution. What is the sum of all possible values of a?

Problem 12

Consider points A, B, C on a circle, with $\overset{\frown}{AB}$ and $\overset{\frown}{BC}$ both quarter circles as shown below.

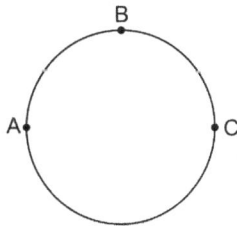

Let a point E be chosen randomly inside the circle. The probability that E is closer to point A than it is to points B and C can be written as $P\%$. What is P? Round your answer to the nearest hundredth if necessary.

Problem 13
Consider real solutions to $(x^2 + 3x + 1)(x^2 + 3x + 4) + 2 = 0$. What is the difference between the largest and smallest solutions? Round your answer to the nearest integer if necessary.

Problem 14
Find the shortest path starting and ending at the origin that goes around the circle $(x-4)^2 + y^2 = 8$. Use the approximation $\pi \approx 3$ and round your answer to the nearest integer.

Problem 15
Let a, b, c, and d be the roots of

$$x^4 + 2018x^3 - 2017x^2 + 2016x - 2015 = 0.$$

The value of

$$\frac{1}{a} + \frac{1}{b} + \frac{1}{c} + \frac{1}{d}$$

can be written as $\dfrac{P}{Q}$ for integers P and Q with $Q > 0$ and $\gcd(P, Q) = 1$. What is $P + Q$?

Problem 16
Suppose we have a rectangle $ABCD$ with $AB = 14$ and $BC = 24$. Draw a circle in the rectangle so that it is tangent to sides $\overline{AB}$, $\overline{BC}$, and $\overline{AD}$. Let M be the midpoint of $\overline{AB}$. Call $E \neq M$ the intersection of $\overline{MD}$ with the circle. Find DE, rounded to the nearest tenth if necessary.

Problem 17
Suppose you have four black, four white, and four green balls. Assume balls of the same color are identical. How many ways are there to put all 12 balls into 4 boxes numbered 1, 2, 3, 4? The balls are NOT in any particular order inside a box.

Problem 18
Let $\lfloor x \rfloor$ represent the greatest integer not exceeding x. Find the sum of all real solutions to $\lfloor x \rfloor^2 - 2x + 1 = 0$. Round your answer to the nearest tenth if necessary.

Problem 19
What is the remainder of $2^{1000} + 3^{1000} + 5^{1000}$ when dividing by 17?

Problem 20
A manufacturer has three production lines that produce a certain kind of IC chip with a failure rate of 1%. Suppose that one of the production lines malfunctions and starts producing IC chips with a failure rate of 10%. Suppose further that chips are randomly packaged and shipped out before the malfunction is noticed. If you get a chip from this lot, the probability it is defective can be written as $\dfrac{N}{M}$ with N, M positive integers and $\gcd(N, M) = 1$. What is $N + M$?

2. ZIML Solutions

This part of the book contains the official solutions to the problems from the nine Junior Varsity ZIML Contests from the 2016-17 School Year.

Students are encouraged to discuss and share their own methods to the problems using the Discussion Forum on ziml.areteem.org.

2.1 ZIML October 2016 Junior Varsity

Below are the solutions from the Junior Varsity ZIML Competition held in October 2016.

The problems from the contest are available on p.15.

Problem 1 Solution
The largest possible number 98765 is not divisible by 9. If we try to change only the last digit, it doesn't work. Hence try the number $\overline{987ab}$. We need $a+b+7+8+9 = a+b+24$ to be divisible by 9, where $a,b \in \{0,1,2,3,4,5,6\}$. From here it is routine to check that $a = 3$, $b = 0$ yields the largest number that works.

Answer: 98730

Problem 2 Solution
There are 8! total ways to line up 8 distinct people. However, since the 3 triplets are identical, we divide by 3! (the number of ways to rearrange them). Hence there are $8! \div 3! = 6720$ different photographs.

Answer: 6720

Problem 3 Solution
Note if $x < 0$, $|x|+x = -x+x = 0$. For $x \geq 0$, $|x|+x = 2x$ so we want $\sqrt{2x} = 8$. This is the same as $2x = 64$ so $x = 32$ is the only solution.

Answer: 32

Problem 4 Solution
The first few terms in the sequence are: $2,5,1,4,3,9,3,9$. From here the sequence always alternates between 3 and 9. More specifically, for $n \geq 5$, $a_n = 3$ if n is odd and $a_n = 9$ if n is even.

Therefore, $a_{100} = 9$.

Answer: 9

Problem 5 Solution

Using Heron's Formula to calculate $[ABC] = 900$. Hence $900 = BC \cdot h/2$ where h is the altitude from $\overline{BC}$. Solve to get $h = 36$.

Answer: 36

Problem 6 Solution

x_1, x_2 are roots of the equation $x^2 + x - 3 = 0$, so $x_1^2 + x_1 - 3 = 0$ and similarly $x_2^2 + x_2 - 3 = 0$. Hence, $x_1^2 = 3 - x_1$, $x_2^2 = 3 - x_2$, and we have

$$x_1^3 - 4x_2^2 + 16 = x_1(3 - x_1) - 4(3 - x_2) + 16.$$

Simplifying and substituting again we have $3x_1 - x_1^2 + 4x_2 + 4 = 3x_1 - (3 - x_1) + 4x_2 + 4 = 4(x_1 + x_2) + 1$. By Viete's Theorem we have $x_1 + x_2 = -1$, so $4(x_1 + x_2) + 1 = 4(-1) + 1 = -3$.

Answer: -3

Problem 7 Solution

Let $\triangle ABC$ be such that $AB = 5$, $BC = 7$, $CA = 8$. Since all three angles add up to $180°$ and all sides have different length, we have that one of the angles is smaller than $60°$ and one is bigger than $60°$. Thus, the $60°$ angle must be $\angle A$. Pick the point D on BC such that $AB = AD$, then $\triangle ABD$ is equilateral, and $\triangle ADC$ is a triangle with sides $(3, 5, 7)$ and $\angle ADC = 120°$, the largest angle in the triangle.

Answer: 120

Problem 8 Solution

Set up the equation $10a + b = 3(a + b)$, so $7a = 2b$. We have $7 \mid b$, so only $b = 7$ works, and then $a = 2$. Thus, the number is

27.

Answer: 27

Problem 9 Solution

Let $P(4) = x$. Then $P(3) = 2x$, $P(2) = 4x$, and $P(1) = 8x$. As all the probabilities sum to 1, we have $x + 2x + 4x + 8x = 15x = 1$. Thus $x = 1/15$. Hence the probability of getting an odd number is $\dfrac{8}{15} + \dfrac{2}{15} = \dfrac{2}{3}$, so $M + N = 5$.

Answer: 5

Problem 10 Solution

Note the equation can be rewritten $3y - 3x = xy$, or after completing the rectangle (or SFFT), $(x-3)(y+3) = -9$. Then $-9 = (1)(-9) = (-1)(9) = (3)(-3) = (-3)(3) = (9)(-1) = (-1)(9)$ leads to 6 solutions to $3y - 3x = xy$. However, one of these is $(0,0)$, which doesn't work as $x, y \neq 0$. The other 5 are $(-6, -2)$, $(2, 6)$, $(4, -12)$, $(6, -6)$, and $(12, -4)$.

Answer: 5

Problem 11 Solution

The shortest distance lies on a line perpendicular to the line given. Such a perpendicular line containing P is $-4x + 3y + 47 = 0$, which intersects the original line at $(34/5, -33/5)$. Using the distance formula this has distance

$$\sqrt{(8 - 34/5)^2 + (-5 + 33/5)^2} = \sqrt{36/25 + 64/25} = \sqrt{4},$$

so $D = 4$.

Answer: 4

Problem 12 Solution
This is a typical stars and bars problem, with 6 stars and $3 - 1 = 2$ bars, where the number of balls in each box is positive. Thus the answer is $\binom{6-1}{2} = 10$.

Answer: 10

Problem 13 Solution
Let $y = (2x^2 - 3x + 1)$. This gives $y^2 = 11y - 10$ so $y^2 - 11y + 10 = 0$ and $(y - 10)(y - 1) = 0$ so $y = 10, y = 1$. Setting $10, 1$ equal to $2x^2 - 3x + 1$ gives $x = -1.5, 0, 1.5, 3$ as our solutions, so the largest and smallest solution differ by 4.5.

Answer: 4.5

Problem 14 Solution
Using the Angle Bisector Theorem, we have

$$AB : AC = BD : DC = 4 : 5.$$

We also know $BC = 9$. Using integer side lengths, the smallest possible values with this ratio are 4 and 5, but this leads to a degenerate triangle. The next smallest values are 8 and 10 which work. Hence the smallest perimeter is $9 + 8 + 10 = 27$.

Answer: 27

Problem 15 Solution
The discriminant of the quadratic equation is $\Delta = (2m - 3)^2 - 4m(m - 3) = 9 > 0$, so the equation has two real roots for all values of m. Using Viete's Theorem, the roots x_1 and x_2 satisfy $x_1 x_2 = m(m - 3)$. Since x_1 and x_2 have opposite signs only when $x_1 x_2 < 0$, we have $m(m - 3) < 0$, which implies $0 < x < 3$. Thus $b - a = 3$.

Answer: 3

Problem 16 Solution

The two chords divide each other in segments of 4 and 8 inches. The perpendicular bisector of each chord goes through the center of the circle, so we can form a right triangle with sides 2 and 6 with hypotenuse r, the radius of the circle. Hence, $r = \sqrt{2^2 + 6^2} = 2\sqrt{10}$, so $A \times B = 2 \times 10 = 20$.

Answer: 20

Problem 17 Solution

Recall that all factors of n come in pairs $a, n/a$, each with product n. The only possible non-paired factor is the square root of a perfect square. Now 60^4 is a perfect square with prime factorization $60^4 = (2^2 \cdot 3 \cdot 5)^4 = 2^8 \cdot 3^4 \cdot 5^4$, so there are $(8+1)(4+1)(4+1) = 225$ factors, with 112 pairs (each multiplying to 60^4) and one square root (which is 60^2). The product of all these factors is $(60^4)^{112} \cdot 60^2 = 600^{450}$.

Answer: 450

Problem 18 Solution

In order for $\triangle ABE$ to be obtuse, E must lie inside the semicircle with diameter AB. This semicircle has area $\pi/2$ and the square has area 4, hence the probability is $\pi/8 \approx 0.3927 \approx 40\%$.

Answer: 40

Problem 19 Solution

Using Vieta's Formulas, $abc + bcd + cda + dab = 2$, and $abcd = -1990$, so $1/a + 1/b + 1/c + 1/d = (abc + bcd + cda + dab)/abcd = -1/995$.

Alternatively: Let $y = 1/x$, then $1 - 2y^3 - 1990y^4 = 0$, and the four roots for y are $1/a, 1/b, 1/c, 1/d$, so according to Vieta's

Formulas, the answer is $\dfrac{-(-2)}{-1990} = -\dfrac{1}{995}$.

Answer: -995

Problem 20 Solution

The total numbers of ways without restriction is $\dbinom{7}{3}$. Whenever the term "at least" is seen, the opposite may be easier to calculate. The opposite is "no pairs of neighbors exist". So we want to choose 3 numbers a, b, c, in increasing order, no neighbors. In order to do that, we need to use "bijection": make a one-to-one correspondence between this problem and a new problem, which has the same answer but easier to count. Let $a' = a, b' = b - 1, c' = c - 2$. So $1 \le a' < b' < c' \le 5$. The numbers a', b', c' do not need to be separated, and the number of ways to choose a', b', c' out of 5 is the same as the number to choose a, b, c out of 7. Thus there are $\dbinom{5}{3}$ ways. Finally, the answer we need is

$$\binom{7}{3} - \binom{5}{3} = 25.$$

Answer: 25

2.2   ZIML November 2016 Junior Varsity

Below are the solutions from the Junior Varsity ZIML Competition held in November 2016.
The problems from the contest are available on p.19.

Problem 1 Solution
Repeated use of the difference between two squares gives

$$a^{32} - b^{32} = (a^{16} + b^{16})(a^8 + b^8)(a^4 + b^4)(a^2 + b^2)(a+b)(a-b).$$

Hence the answer is 6.

Answer: 6

Problem 2 Solution
Group by one-digit, two-digit, etc., numbers. Thus you would write

$$1 \cdot 9 + 2 \cdot 90 + 3 \cdot 900 + 4 \cdot 9000 + 5 \cdot 1 = 38894.$$

digits in total.

Answer: 38894

Problem 3 Solution
Since $g(u) = u^2 - 13u + 12 = (u-1)(u-12)$ is a parabola opening upwards with roots at $u = 1, 12$, $g(u)$ is negative for $1 < u < 12$. Substituting $u = x^2$ we have $f(x)$ is negative when $1 < x^2 < 12$ or $1 < |x| < 2\sqrt{3}$. Since $2\sqrt{3} \approx 3.46$, $f(n)$ is negative for $n = \pm 2, \pm 3$, so the answer is 4.

Answer: 4

Problem 4 Solution
Only numbers that end in 1, 5 or 6 end with the same digit when squared. A number that ends in 1 is of the form $10a + 1$, and

$$(10a + 1)^2 = 120a + 1.$$

Note the tens digit of this square number will be the same as the ones digit of $2a$, so the only number that would work is $a = 0$, which does not yield a 2-digit number. A number that ends in 5 is of the form $10a + 5$, and

$$(10a + 5)^2 = 200a + 25,$$

that is, it always ends in 25; so 25 is the only 2-digit number ending in 5 that ends with the same two digits when squared. A number that ends in 6 is of the form $10a + 6$, and

$$(10a + 6)^2 = 220a + 36.$$

Note the tens digit of this square number will be 3 more than the ones digit of $2a$. Noticing this pattern, we see the only number that works is $a = 7$, so 76 is the only other 2-digit number that ends with the same two digits when squared. Thus, the answer is $25 + 76 = 101$.

Answer: 101

Problem 5 Solution
Note the path below (1 to 2 to 3 ... to 8 to 1) is on the surface of the cube and is of length $2 + 6\sqrt{2}$.

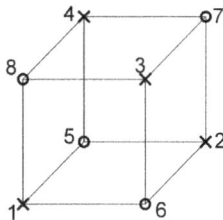

To show this is optimal, consider the points as labeled with either X or O. Note traveling from X to X (when possible) is of length $\sqrt{2}$ and the same fact holds true for O to O. Traveling from X to

O or O to X (when possible) is of length 1. Since the path starts and ends at and X we must have at least 2 of the 8 steps of length 1. This implies that $2 + 6\sqrt{2}$ is the longest path.

Answer: 6

Problem 6 Solution

If a and b are both odd primes, $p = a^b + b^a$ is an even number greater than 2, not a prime. So a or b should be 2. Assume $a = 2$ without loss of generality, and let $b = 2k + 1$. So $p = 2^{2k+1} + (2k+1)^2 = 2 \cdot 4^k + b^2$. If $b \neq 3$, $b^2 \equiv 1 \pmod 3$, and so $2 \cdot 4^k + b^2 \equiv 2 + 1 \equiv 0 \pmod 3$, and clearly $p > 3$, so p is not a prime, a contradiction. Therefore $b = 3$. At the end, $p = a^b + b^a = 2^3 + 3^2 = 17$.

Answer: 17

Problem 7 Solution

Note using the triangle inequality, we must have $a + b > 1$, $b + 1 > a$, $a + 1 > b$. Graphing where all 3 of these inequalities is true yields the following (restricted to $0 \leq a, b \leq 2$)

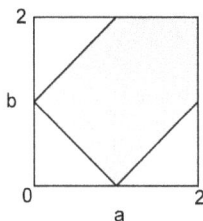

The entire square has area 4, and the shaded region has area $\dfrac{5}{2}$, so the probability is

$$\frac{5/2}{4} = \frac{5}{8} = 0.625 = 62.5\%.$$

Answer: 62.5

Problem 8 Solution

Work starting with the largest possible number 987654321. Trial and error tells us we must change at least the last four digits, so suppose the number is $98765dcba$ with a, b, c, d chosen from $1, 2, 3, 4$. Using the divisibility rule for 11, we want $a - b + c - d + 5 - 6 + 7 - 8 + 9 = a - b + c - d + 7$ to be a multiple of 11. The only possibility that works is if $a - b = c - d = 2$. We choose $a = 3, b = 1, c = 4, d = 2$ to give the largest possible number: 987652413.

Answer: 987652413

Problem 9 Solution

First note $\overline{CD} \parallel \overline{AB}$ as the arcs $\overset{\frown}{AC}$ and $\overset{\frown}{DB}$ have the same measure. Therefore $[ACD] = [OCD]$ where O is the center of the circle. Hence the area of the figure we want is the same as the area of the sector with arc $\overset{\frown}{CD}$. Since $\overset{\frown}{CD}$ is $\frac{2}{3}$ of arc $\overset{\frown}{AB}$ it is a one-third of a circle with radius 6. Hence the area we want is

$$\frac{1}{3} \cdot \pi \cdot 6^2 = 12\pi,$$

so $K = 12$.

Answer: 12

Problem 10 Solution

The discriminant of the quadratic equation is $\Delta = (3m + 2)^2 - 4m(3 - m) = 13m^2 + 4 > 0$, so the equation has two real roots for all values of m. Using Vieta's Theorem, the roots x_1 and x_2 satisfy $x_1 x_2 = m(3 - m)$. Since x_1 and x_2 have opposite signs only when $x_1 x_2 < 0$, we have $m(3 - m) < 0$, which implies $0 < m < 3$. The integers in this interval are 1 and 2 so the sum is $1 + 2 = 3$.

Answer: 3

Problem 11 Solution

Since 11 is prime, we can use Fermat's Little Theorem to get $4^{10} \equiv 1 \pmod{11}$. Therefore, $4^{2016} \equiv 4^6 \equiv 2^{12} \pmod{11}$. Again Fermat's Little Theorem says $2^{10} \equiv 1 \pmod{11}$, so $2^{12} \equiv 2^2 \equiv 4 \pmod{11}$ so our answer is 4.

Answer: 4

Problem 12 Solution

a and b are the roots of an equation $(x+1)^2 + 3(x+1) - 3 = 0$, which is $x^2 + 5x + 1 = 0$. The discriminant is positive. We have: $a + b = -5$, and $ab = 1$. Thus $a, b < 0$ and

$$b\sqrt{\frac{b}{a}} + a\sqrt{\frac{a}{b}} = -\frac{b}{a}\sqrt{ab} - \frac{a}{b}\sqrt{ab}$$

$$= -\frac{a^2 + b^2}{ab}\sqrt{ab}$$

$$= -\frac{(a+b)^2 - 2ab}{\sqrt{ab}}$$

$$= -23.$$

Answer: -23

Problem 13 Solution

Let E be midpoint of $\overline{AD}$, connect $\overline{EO}$ and extend and intersect $\overline{BC}$ at F. Then $\overline{EO} \perp \overline{AD}$, so $\overline{EO} \perp \overline{BC}$, thus F is the midpoint of $\overline{BC}$. Since $OA = 10$, and $AE = 12/2 = 6$, we get $OE = 8$. Similarly, $OB = 10$ and $BF = 16/2 = 8$, so we get $OF = 6$. So $EF = 8 + 6 = 14$. Then

$$[ABCD] = \frac{(AD + BC) \cdot EF}{2} = \frac{(12 + 16) \cdot 14}{2} = 196.$$

Answer: 196

Problem 14 Solution

Factoring the number into primes, each pair of 2 and 5 make a 0 at the end. There are more 2's than 5's so we only need to count the factor 5. From 1 to 1000 there are 200 multiples of 5 (each contributing one 5), 40 multiples of $25 = 5^2$ (each contributing an extra 5), 8 multiples of $125 = 5^3$ (contributing one more extra 5), and lastly 1 multiple of 625 (contributing a last extra 5), for a total of $200 + 40 + 8 + 1 = 249$.

Answer: 249

Problem 15 Solution

Using stars and bars, there are

$$\binom{8}{2} = 28$$

positive solutions to $a + b + c = 9$. However, $1, 1, 7$ (and its rearrangements) are not possible here, so we subtract 3. Hence there are $28 - 3 = 25$ ways to sum to 9. As there are $6^3 = 216$ total outcomes, the probability is $\dfrac{25}{216}$ so $P + Q = 25 + 216 = 241$.

Answer: 241

Problem 16 Solution

Using the angle bisector theorem we have $AB : AC = BE : EC = 2 : 3$ so $AB = 8$. Similarly we find $AD = 15$. Hence the perimeter is $8 + 10 + 18 + 15 = 51$.

Answer: 51

Problem 17 Solution

Note we can factor/rewrite the equation as

$$\frac{x^2 + xy + y^2}{x^3 y^2 + x^2 y^3 + x + y} = \frac{(x+y)^2 - xy}{((xy)^2 + 1)(x+y)}.$$

Since $x+y=6$ and $xy=9-5=4$, we can then substitute to get that the expression is equal to

$$\frac{36-4}{17\cdot 6} = \frac{16}{51}.$$

Hence $N+M = 16+51 = 67$.

Answer: 67

Problem 18 Solution
Mod 3, the numbers are $0,1,2,0,1$. Arranging the numbers mod 3, the only two possibilities are $1,0,2,1,0$ or $0,1,2,0,1$. For each of these possibilities there are $2! = 2$ ways to arrange 15 and 45 (the 0 mod 3 numbers) and $2! = 2$ ways to arrange 25 and 55 (the 1 mod 3 numbers). Hence there are $2^3 = 8$ arrangements in total.

Answer: 8

Problem 19 Solution
Consider the following diagram:

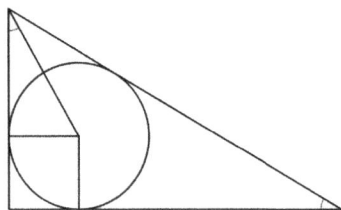

The two marked angles are both $30°$ and the quadrilateral in the bottom left is a square. Hence the short side of the triangle is $1+\sqrt{3}$ and the hypotenuse is $2(1+\sqrt{3})$. Since this is a right triangle, this is the diameter of the larger circle, so the radius is $1+\sqrt{3}$. Thus the larger circle has area

$$\pi(1+\sqrt{3})^2 = (4+2\sqrt{3})\pi,$$

which is $4 + 2\sqrt{3} = 4 + \sqrt{12}$ times larger than the original circle of area π. Therefore our answer is $4 + 12 = 16$.

Answer: 16

Problem 20 Solution

We can use the Principal of Inclusion-Exclusion (PIE) to calculate the number of ways for them to leave so that at least one person has their own coat, and subtract it from the $4! = 24$ total outcomes. Let A, B, C, D be be the event the 1st, 2nd, 3rd, and 4th person (respectively) get their own coat. Then, using PIE we have

$$
\begin{aligned}
n(A \cup B \cup C \cup D) &= \binom{4}{1} \cdot 3! - \binom{4}{2} \cdot 2! + \binom{4}{3} \cdot 1! - \binom{4}{4} \\
&= 4! - \frac{4!}{2!} + \frac{4!}{3!} - \frac{4!}{4!} \\
&= 15.
\end{aligned}
$$

Hence our final answer is $24 - 15 = 9$.

Answer: 9

2.3   ZIML December 2016 Junior Varsity

Below are the solutions from the Junior Varsity ZIML Competition held in December 2016.
The problems from the contest are available on p.25.

Problem 1 Solution
There are 3 choices for each of his four sons (total of $3^4 = 81$ possibilities). For the daughters, there are $4 \cdot 3 \cdot 2 = 24$ possibilities since they all must be at a different school. The final answer is $81 \cdot 24 = 1944$.

Answer: 1944

Problem 2 Solution
Regardless of how many of the terms inside the absolute values are negative, we either get (i)

$$\frac{x+4}{x+1} = \frac{x+3}{x+2}$$

or (ii)

$$\frac{x+4}{x+1} = -\frac{x+3}{x+2}.$$

In case (i) we get

$$(x+4)(x+2) = (x+1)(x+3),$$

then

$$x^2 + 6x + 8 = x^2 + 4x + 3,$$

and

$$x = \frac{-5}{2}.$$

In case (ii) we get

$$(x+4)(x+2) = -(x+1)(x+3),$$

then
$$x^2 + 6x + 8 = -x^2 - 4x - 3,$$

so
$$2x^2 + 10x + 11 = 0,$$

and
$$x = \frac{-5}{2} \pm \frac{\sqrt{3}}{2},$$

using the quadratic equation. Hence the only rational solution is $\frac{-5}{2}$, so $P + Q = -5 + 2 = -3$.

Answer: -3

Problem 3 Solution

We are given $[ABCD] = 30$ (the area is 30). Note $[ABD] = [DEA]$ as they have the same height and base. Hence, after removing the shared $\triangle AFD$ we see the two shaded triangles have the same area. Then (since E is a midpoint) note $[DEC] = [BED] = [ABCD]/4$. Since $ABCD$ is a parallelogram and $BE = AD/2$, $\triangle AFD \sim \triangle EFB$, with ratio of side lengths $2:1$. Using this information, we have $[DFE] = 2[BEF]$. Therefore, $[DFE] = [ABF] = [ABCD]/6$. Hence the sum of the shaded regions is $[ABCD]/3 = 30/3 = 10$.

Answer: 10

Problem 4 Solution

$36 = 4 \cdot 9$, so we check divisibility by 4 and 9. Divisibility by 4 means the last two digits must be divisible by 4, and there are two cases for y: 2 or 6. For each of these cases, x is determined by the divisibility rule for 9. This gives either 11232 or 61236. The difference is $61236 - 11232 = 50004$.

Answer: 50004

Problem 5 Solution

Since AE is the angle bisector, we have $AB:AC=BE:EC=2:5$. Let $AB=2x$ and $BC=5x$. Then $2x+5x+4+10=35$ so solving for x we have $x=3$. Hence $AC=5x=5\cdot 3=15$.

Answer: 15

Problem 6 Solution

First note using Fermat's Little Theorem, $3^{10}\equiv 1\pmod{11}$. Further, using the extension (or patterns), $3^4\equiv 1\pmod{10}$. Therefore,

$$3^3+3^{3^2}+3^{3^3}+3^{3^4}\equiv 3^{3^5}+\cdots+3^{3^8}\pmod{11}.$$

Hence

$$3^3+3^{3^2}+3^{3^3}+\cdots+3^{3^8}\equiv 2\cdot(3^3+3^{3^2}+3^{3^3}+3^{3^4})\pmod{11}.$$

Using patterns we have $3^3\equiv 5\pmod{11}$, $3^{3^2}\equiv 3^9\equiv 4\pmod{11}$, $3^{3^3}\equiv 3^7\equiv 9\pmod{11}$, and $3^{3^4}\equiv 3^1\equiv 3\pmod{11}$. Thus, the final answer is $2\cdot(5+4+9+3)\equiv 9\pmod{11}$.

Answer: 9

Problem 7 Solution

Let $y=\sqrt{x^2-2x+5}$ so the equation becomes $4y-4\sqrt{y^2-1}=1$. Solving this for y gives $y=\dfrac{17}{8}$. Hence we need to solve

$$\frac{289}{64}=x^2-2x+5.$$

The solutions are

$$1\pm\frac{\sqrt{33}}{8}.$$

Hence the sum is 2. (Alternatively you can use Viete's on the quadratic here.)

Answer: 2

Problem 8 Solution

The sum of the two dice is 6, so the possibilities for the two rolls are:

$$(1,5),(2,4),(3,3),(4,2),(5,1).$$

As two of these possibilities have a 2, the probability is

$$\frac{2}{5} = 40\%$$

so $K = 40$.

Answer: 40

Problem 9 Solution

We have $C = (1,1)$ and solving we get $A = (1,3)$, $B = (3,1)$. Hence $\triangle ABC$ is a right triangle with side lengths 2, so the area of $\triangle ABC$ is $2^2/2 = 2$.

Answer: 2

Problem 10 Solution

Circle ω_1 has center $A(-6,8)$ and radius 4, and circle ω_2 has center $B(9,-12)$ and radius 6. The length

$$AB = \sqrt{(9+6)^2 + (12+8)^2} = 25,$$

and the two circles are separate.

For two separate circles, the length of the common *external* tangent line segment is greater than the length of the common *internal* tangent line segment (*Note to reader: can you prove this fact?*). Thus, the segment $\overline{PQ}$ is the common external tangent line segment to ω_1 and ω_2. From the problem, P is on ω_1 and Q is on ω_2.

Let R be the point on $\overline{PA}$ such that $PR = QB$. Then $PQBR$ is a rectangle, and $\triangle ARB$ is a right triangle. We know that $AB = 25$,

---

$AR = AP - PR = 6 - 4 = 2$, thus $BR = \sqrt{25^2 - 2^2} = \sqrt{621}$, and so $PQ = BR = \sqrt{621}$, therefore $M = 621$.

Answer: 621

Problem 11 Solution

Consider the following cases for prime number z: (1) If $z = 2$, $x(x + y) = 122 = 2 \times 61$. Thus $x = 2, y = 59$. Then $x = z$ which is not what we want. (2) If z is an odd prime, then $x(x + y)$ is an odd number. That means both x and $x + y$ are odd, and so $y = 2$. The equation becomes $x(x + 2) = z + 120$, and so $x^2 + 2x - 120 = z$, so $(x + 12)(x - 10) = z$. But z is prime, then $x - 10 = 1$, so $x = 11, z = 23$. Hence the triple is $(11, 2, 23)$ so $x + y + z = 11 + 2 + 23 = 36$.

Answer: 36

Problem 12 Solution

Let a, b, c stand for the time Alice, Bob, and Charlie arrive for dinner, measured as a fraction of an hour after 5 PM. They will have dinner if $a < b < c$. We can then view their arrival times as a coordinate (a, b, c). The inequality $a < b < c$ is a triangular prism with base area $1/2$ and height 1 (so volume $1/6$). As their arrival times form a cube of volume 1, the probability is $1/6$. Hence $M - N = 6 - 1 = 5$.

Answer: 5

Problem 13 Solution

The factors come in pairs, each pair having a product of n. So, the factors 1 and n form one pair, and if n is not one, there will be another pair. That means $n = 1$ or n has exactly 4 factors, that is, the product of two primes (pq) or the cube of a prime (p^3). The possible numbers are: 1, 6, 8, 10, 14, 15, 21, 22, 26, and 27, a total of 10 possibilities.

Answer: 10

Problem 14 Solution

Label the vertices of the cube $ABCD - A'B'C'D'$ (with A' above A, B' above B, etc.) If 3 vertices are contained in the same square face, then we have a tetrahedron (which is a triangular pyramid) with base of area 18 and height of 6. Thus, the volume is

$$\frac{1}{3} \cdot 18 \cdot 6 = 36.$$

Now suppose we have tetrahedron $AC - B'D'$. Note this tetrahedron divides the rest of the cube into 4 congruent tetrahedra, which we just saw have volume 36. Hence, tetrahedron $AC - B'D'$ has volume $6^3 - 4 \cdot 36 = 2 \cdot 36 = 72$.

Answer: 72

Problem 15 Solution

Note that using $1, 2, 4, 8, 16, 32$ (the exponents of x, which are all powers of 2) the only way to get a sum of 53 is $53 = 1 + 4 + 16 + 32$. (This is because, converted to binary, $53_{10} = 110101_2$.) Thus the power $x^{53} = x^{32} \cdot x^{16} \cdot x^4 \cdot x^1$, and the coefficient is the product of the constant terms whose power of x does not appear, so the answer is $8 \cdot 9 = 72$.

Answer: 72

Problem 16 Solution

Let $u = x + y$, $v = xy$, then $u^2 - 3v = 13$, and $u - v = -5$. Solve to get $(u, v) = (7, 12)$ or $(u, v) = (-4, 1)$. For each pair of (u, v), set up quadratic equation in t: $t^2 - ut + v = 0$ and x, y are the two roots. So $t^2 - 7t + 12 = 0$, or $t^2 + 4t + 1 = 0$. The solutions for (x, y) are $(3, 4), (4, 3), (-2 + \sqrt{3}, -2 - \sqrt{3}), (-2 - \sqrt{3}, -2 +$

$\sqrt{3}$).

Answer: 4

Problem 17 Solution

Arrange the red cards (only 1 way). Then place 2 black cards in between the cards (only 1 way) using $2 \cdot 7 = 14$ black cards. Then the remaining 2 cards can arranged in the 9 spaces created by the red cards (now including the endpoints) using stars and bars. Hence our answer is

$$\binom{2+9-1}{2} = \binom{10}{2} = 45.$$

Answer: 45

Problem 18 Solution

Using the quadratic formula, we find that

$$|x_1 - x_2| = \frac{\sqrt{m^2 + 64}}{2} = m - 1.$$

Squaring and simplifying leads to the quadratic equation

$$3m^2 - 8m - 60 = (3m + 10)(m - 6) = 0,$$

so $m = -\dfrac{10}{3}$ or 6. Because we squared both sides of the equation for $x_1 - x_2$ to get rid of the radical, we need to check for extraneous solutions. Since $\dfrac{\sqrt{(-10/3)^2 + 64}}{2} > 0$ and $-\dfrac{10}{3} - 1 < 0$, we reject the solution $m = -\dfrac{10}{3}$. The solution $m = 6$ checks out fine. Hence the only solution is $m = 6$, and our answer is 6.

Answer: 6

Problem 19 Solution

Let X, Y, Z, be, respectively, the set of students that get an A on the first exam, the second exam, and the third exam. Then by the Principle of Inclusion-Exclusion (PIE),

$$
\begin{aligned}
n(X \cap Y \cap Z) &= n(X \cup Y \cup Z) - n(X) - n(Y) - n(Z) \\
&\quad + n(X \cap Y) + n(X \cap Z) + n(Y \cap Z) \\
&= 15 - 3 \cdot 10 + 3 \cdot 8 \\
&= 9.
\end{aligned}
$$

Answer: 9

Problem 20 Solution

$\overline{abcd} = \overline{ab} \times 100 + \overline{cd} = \overline{ab} \times 99 + \overline{ab} + \overline{cd}$, so in fact the requirement is simply that the original four-digit number $\overline{abcd}$ is divisible by 11. The number of four-digit numbers divisible by 11 is $\lfloor 9999/11 \rfloor = 909$.

Answer: 909

2.4   ZIML January 2017 Junior Varsity

Below are the solutions from the Junior Varsity ZIML Competition held in January 2017.

The problems from the contest are available on p.31.

Problem 1 Solution

Let $x = AB$. Then as $\triangle ABD$ is a 45-45-90 triangle, so $DB = x$ as well. As $\triangle ABC$ is a 30-60-90 triangle, $BC = \sqrt{3}AB$ so $x + 2 = x\sqrt{3}$. Hence

$$x = \frac{2}{\sqrt{3} - 1} = \sqrt{3} + 1 \approx 3.$$

Answer: 3

Problem 2 Solution

It can be determined immediately that $e = 5$. Also we know that b, d, f must be even digits, and so a, c are odd. So either $a = 1, c = 3$ or $c = 1, a = 3$. To make $\overline{abc}$ a multiple of 3, only $b = 2$ works. To determine d, note that we only need to have $4 \mid \overline{cd}$, and it could be 16 or 36, but not 14 or 34. So $d = 6$ and then $f = 4$. Hence 123654 and 321654 are the two possible solutions, with difference 198000.

Answer: 198000

Problem 3 Solution

The first rook can be on any of the 64 squares, and this rook removes $8 + 8 - 1 = 15$ squares for the second rook. We divide by 2! as the rooks are identical. Hence there are

$$64 \cdot 49 / 2! = 1568$$

different arrangements.

Answer: 1568

Problem 4 Solution

The shortest path can be determined as follows: let point C be the midpoint of one of the edges of the middle cube.

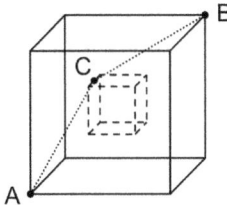

The length of AC is equivalent to determining the length of the longest diagonal of a $1 \times \frac{3}{2} \times 2$ rectangular prism and the length of CB is equivalent to determining the length of the longest diagonal of a $2 \times \frac{3}{2} \times 1$ rectangular prism. Therefore,

$$AC = \sqrt{1^2 + \left(\frac{3}{2}\right)^2 + 2^2} = \frac{\sqrt{29}}{2}$$

$$CB = \sqrt{2^2 + \left(\frac{3}{2}\right)^2 + 1^2} = \frac{\sqrt{29}}{2}$$

The shortest length of a path from point A to B is $\sqrt{29}$.

Answer: 29

Problem 5 Solution

Let $y = 2x^2 - 3$, then $\dfrac{1}{y} - 4y = 0$. Solve and get $y = \pm\dfrac{1}{2}$. Then solve for x to get $x = \pm\dfrac{\sqrt{7}}{2}$ and $x = \pm\dfrac{\sqrt{5}}{2}$. All are verified to be solutions, and the largest is

$$\frac{\sqrt{7}}{2} = \sqrt{\frac{7}{4}},$$

so $A + B = 7 + 4 = 11$.

Answer: 11

Problem 6 Solution

It is easy to calculate that ℓ_3 intersects ℓ_1 and ℓ_2 at $(4, 2)$ and $(2, 4)$, respectively, and the midpoint of these is $(3, 3)$. Note that $\ell_1 \| \ell_2$, so if we have a circle at $(3, 3)$ that is tangent to ℓ_1 it is also tangent to ℓ_2. Therefore, the largest circle must have center $(3, 3)$ and radius $\sqrt{2}$. Hence it has equation

$$(x - 3)^2 + (y - 3)^2 = 2, \quad \text{or} \quad x^2 + y^2 - 6x - 6y + 16 = 0,$$

so $C = 16$.

Answer: 16

Problem 7 Solution

Let $m = 2121 \cdots 212$ be k pairs of 21 followed by a 2. Then $m \equiv k(2 - 1) + 2 \equiv k + 2 \pmod{11}$. Hence $n \equiv 2 + 3 + 4 + 5 + 6 + 7 \equiv 5 \pmod{11}$.

Answer: 5

Problem 8 Solution

There are $5!$ ways to arrange the consonants. This creates $5 + 1 = 6$ 'spaces' that can be used for placing the vowels. Hence there are

$$5! \cdot 6 \cdot 5 \cdot 4 = 14400$$

total arrangements.

Answer: 14400

Problem 9 Solution

First note if $a \geq 4$, then

$$\frac{1}{a} + \frac{1}{b} + \frac{1}{c} \leq \frac{3}{4} < 1$$

so we must have $a = 1$, 2, or 3. Working through the cases we see the only solutions are

$$(2,3,6),(2,4,4),(3,3,3)$$

so there are 3 such solutions.

Answer: 3

Problem 10 Solution

Let $z = x^2 + 4x + 8$. Then our equation becomes

$$z^2 + 3xz + 2x^2 = 0.$$

The left hand side can be factored as

$$z^2 + 3xz + 2x^2 = (z + 2x)(z + x).$$

Re-substituting we have

$$\begin{aligned}
&(x^2 + 4x + 8 + 2x)(x^2 + 4x + 8 + x)\\
&= (x^2 + 6x + 8)(x^2 + 5x + 8)\\
&= (x + 2)(x + 4)(x^2 + 5x + 8).
\end{aligned}$$

Hence the solutions to the equation are $x = -2$ and $x = -4$, as $x^2 + 5x + 8 = 0$ does not have real solutions. Therefore, there are 2 rational solutions.

Answer: 2

Problem 11 Solution

Start with the first number: it is either 1 or 10 (2 choices). For the second number, it is either the largest remaining, or the smallest remaining (2 choices again). This pattern continues for all the numbers except the 10th. There are thus $2^9 = 512$ permutations.

Answer: 512

Problem 12 Solution

Note a, b, and 1 are solutions to $x^3 + 4x^2 - 9x + 4 = 0$. Let $c = 1$. Then using Viete's theorem

$$\frac{1}{a} + \frac{1}{b} + \frac{1}{c} = \frac{bc + ac + ab}{abc} = \frac{-9}{-4} = \frac{9}{4}$$

Since $\dfrac{1}{c} = 1$, we have $\dfrac{1}{a} + \dfrac{1}{b} = \dfrac{9}{4} - 1 = \dfrac{5}{4}$. Hence $Q - P = 4 - 5 = -1$.

Answer: -1

Problem 13 Solution

We want $2^n \equiv 1 \pmod{31}$. Note $2^5 \equiv 1 \pmod{31}$, so consider $n = 5k + l$ for integers k and $0 \le l < 5$. Then $2^n \equiv 2^l \pmod{31}$ and for such l, the only possibility is $l = 0$. Hence $2^n - 1$ is divisible by 31 if and only if n is a multiple of 5, and there are 20 such positive integers less than or equal to 100.

Answer: 20

Problem 14 Solution

Using the discriminant,

$$\Delta = -4(2a^2 - 2a + 4b^2 + 4ab + 1)$$
$$= -4[(a-1)^2 + (a+2b)^2] \ge 0,$$

so must have $a = 1$ and $b = -\dfrac{1}{2}$. Hence $a + b = 0.5$.

Answer: 0.5

Problem 15 Solution

Let $N = \overline{abc}$ be a three digit positive number where a, b, c represent the digits of N. Note that for N to have digits that are in an arithmetic progression, we need

$$b = \frac{a+c}{2}.$$

Since b is a digit of N, a and c must both be odd or both be even. In the case when a and c are both odd, there are

$$5 \times 5 = 25$$

possibilities of this happening. In the case when a and c are both even, there are

$$4 \times 5 = 20$$

possibilities of this happening since a cannot be 0. Therefore, there are a total of

$$25 + 20 = 45$$

possible N such that its digits are in an arithmetic progression.

There are a total of

$$9 \times 10 \times 10 = 900$$

possible three digit positive numbers. Therefore, the probability that its digits are in an arithmetic progression is

$$\frac{45}{900} = \frac{1}{20} = 5\%.$$

Thus $P = 5$.

Answer: 5

Problem 16 Solution
Note by power of a point, $PR \cdot QR = RT^2$ so

$$RT^2 = (12 + 4) \cdot 4 = 64,$$

so $RT = 8$.

Answer: 8

Problem 17 Solution

Divide the set $\{1,2,\ldots,100\}$ into 2 categories: the first category consists of numbers that are multiples of 3, and the second category consists of all other numbers. The number of numbers in the first category is $\left\lfloor \dfrac{100}{3} \right\rfloor = 33$. Two numbers have common divisor 3 only when they are both in the first category. There are $100 - 33 = 67$ numbers in the second category, so if we pick 69 numbers, we're guaranteed to have at least two numbers with common divisor 3.

Answer: 69

Problem 18 Solution

Denote the events of choosing a line by L_1, L_2, and L_3, with L_1 being the malfunctioning line, and let A be the event that your chip is defective. Then, by the law of total probability,

$$P(A) = P(A|L_1)P(L_1) + P(A|L_1^c)P(L_1^c) = \frac{1}{10} \cdot \frac{1}{3} + \frac{1}{100} \cdot \frac{2}{3} = \frac{1}{25}.$$

Hence $N + M = 1 + 25 = 26$.

Answer: 26

Problem 19 Solution

Say the medians are $\overline{AD}$, $\overline{BE}$, $\overline{CF}$, which divide $\triangle ABC$ into six equal area triangles. We know $GD = 5/2$, $GE = 6$, and $GF = 13/2$, as G is the centroid. Extend $\overline{BE}$ 6 units further from E, giving a point H.

Note $AGCH$ is a parallelogram (its diagonals intersect in their midpoints), so $CH = 5$. Hence, $\triangle GCH$ is a right triangle with area 30. Note that $[GCH] = [ACG] = [ABC]/3$ and thus $[ABC] = 90$.

Answer: 90

Problem 20 Solution

Squaring both sides we have $2x^2 + 4 = x^2 + 2kx + k^2$ so $x^2 - 2kx + 4 - k^2 = 0$. The discriminant of this quadratic equation is

$$\Delta = 4k^2 - 4(4 - k^2) = 8k^2 - 16.$$

Hence the equation has real roots when $8k^2 \geq 16$ so $k^2 \geq 2$, and thus $k \geq \sqrt{2}$ or $k \leq -\sqrt{2}$. However, if $k \leq -\sqrt{2}$ we have extraneous roots, so $k \geq \sqrt{2}$. (It may help to think of this part graphically.) Since k must be an integer with $|k| \leq 5$, there are 4 possible values for k ($k = 2, 3, 4, 5$).

Answer: 4

2.5 ZIML February 2017 Junior Varsity

Below are the solutions from the Junior Varsity ZIML Competition held in February 2017.

The problems from the contest are available on p.37.

Problem 1 Solution

First note we can rewrite the expression as

$$\frac{x^8 - 1}{x - 1}.$$

Using differences of two squares the numerator is

$$(x-1)(x+1)(x^2+1)(x^4+1),$$

so after canceling we have $(x+1)(x^2+1)(x^4+1)$, which contains 3 binomials.

Answer: 3

Problem 2 Solution

Consider the following general diagram:

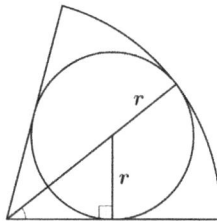

In this case, the sector is $60°$, so the marked angle is $30°$ and the triangle is a 30-60-90 triangle with hypotenuse $2r$. Hence, $12 = r + 2r$ so solving for r gives $r = 4$.

Answer: 4

Problem 3 Solution

Note that

$$\overline{a0}^2 + \overline{a1}^2 + \overline{a2}^2 + \cdots + \overline{a9}^2 \equiv 1^2 + 2^2 + \cdots + 9^2 \pmod{10}$$

(for example, $20^2 + 21^2 + \cdots 29^2 \equiv 0^2 + 1^2 + \cdots + 9^2$). Hence, $1^2 + 2^2 + \cdots 99^2 \equiv 10 \times (1^2 + 2^2 + \cdots + 9^2) \equiv 0 \pmod{10}$ so the units digit is 0.

Answer: 0

Problem 4 Solution

You want to buy $6 \div 2 = 3$ pairs of cookies. We can use stars and bars: with 3 stars and, as there are 8 choices for cookies, $8 - 1 = 7$ bars. Therefore, there are

$$\binom{3+8-1}{3} = 120$$

ways of buying the cookies.

Answer: 120

Problem 5 Solution

x_1 and x_2 are roots of the equation $x^2 + x - 4 = 0$, so $x_1^2 + x_1 - 4 = 0$ and $x_2^2 + x_2 - 4 = 0$. Hence, $x_1^2 = 4 - x_1$, $x_2^2 = 4 - x_2$, and we have

$$
\begin{aligned}
x_1^3 - 5x_2^2 + 25 &= x_1(4 - x_1) - 5(4 - x_2) + 25 \\
&= 4x_1 - x_1^2 - 20 + 5x_2 + 25 \\
&= 4x_1 - (4 - x_1) + 5x_2 + 5 \\
&= 5x_1 + 5x_2 + 1.
\end{aligned}
$$

By Viete's Theorem we have $x_1 + x_2 = -1$, so $5x_1 + 5x_2 + 1 = 5(-1) + 1 = 1$.

Answer: 1

Problem 6 Solution

Note the smallest possible number, 30124, does not work (as $4-2+1-0+3=6$), but changing the 4 to a 9 (yielding the number 30129) does work.

Answer: 30129

Problem 7 Solution

The prime factorizations of 108 and 264 are $2^2 \cdot 7^2$ and $2^3 \cdot 2 \cdot 11$, respectively. So, their lcm is $2^3 \cdot 3 \cdot 7^2 \cdot 11$. The smallest multiple of this lcm that is a perfect square is $2^4 \cdot 3^2 \cdot 7^2 \cdot 11^2 = 853776$.

Answer: 853776

Problem 8 Solution

We will use PIE. Let A represent the ways we can arrange the numbers so that 2 is next to 1, and B the ways we can arrange the numbers so that 2 is next to 3. Then $n(A) = n(B) = 2! \cdot 4! = 48$ and $n(A \cap B) = 2! \cdot 3! = 12$. Thus, there are $48 + 48 - 12 = 84$ ways of doing this.

Answer: 84

Problem 9 Solution

Note using the Pythagorean Theorem, $AB = 13$. If $x = DC$, then $BD = 12 - x$. Using the Angle Bisector Theorem,

$$\frac{13}{5} = \frac{AB}{AC} = \frac{BD}{DC} = \frac{12-x}{x}.$$

Solving for x we have

$$x = \frac{10}{3} \Rightarrow 12 - x = \frac{26}{3}.$$

Therefore, the area of $\triangle ABD$ is

$$\frac{65}{3} = 21.\overline{6} \approx 22.$$

Answer: 22

Problem 10 Solution

Let A be the event that the rolls add up to 6, and T the event that you get tails. Then $P(T) = P(T^c) = 1/2$, $P(A|T) = 5/36$, and $P(A|T^c) = 1/6$. Using Bayes' Theorem we have that

$$P(T|A) = \frac{5/72}{(5/72) + (1/12)} = \frac{5}{11}.$$

Hence $N + M = 5 + 11 = 16$.

Answer: 16

Problem 11 Solution

By Wilson's Theorem we have $28! \equiv -1 \pmod{29}$. Note that $28! \equiv 28 \cdot 27 \cdot 26! \equiv 2 \cdot 26! \pmod{29}$. Therefore, $2 \cdot 26! \equiv -1 \pmod{29}$, since $2 \cdot 15 \equiv 1 \pmod{29}$ we have

$$26! \equiv 15 \cdot (-1) \equiv 14 \pmod{29}.$$

Answer: 14

Problem 12 Solution

Expanding the right side we get $|x^2 + 6x + 5|$, so with the substitution $y = x^2 + 6x + 1$ we get $|y| = |y + 4|$. Hence we must have $y = -y - 4$, so $y = -2$. Thus we need $x^2 + 6x + 1 = -2$ or $x^2 + 6x + 3 = 0$. The quadratic formula gives us $-3 \pm \sqrt{6}$ as roots. Hence $A \cdot B = -3 \cdot 6 = -18$.

Answer: -18

Problem 13 Solution

Note if $x < 0$, $|x| + 2x = -x + 2x = x < 0$. For $x \geq 0$, $|x| + 2x = 3x$ so we want $\sqrt{3x} = 12$. This is the same as $3x = 144$ so $x = 48$ is the only solution.

Answer: 48

Problem 14 Solution

First note that there are the same number of ways to choose either committee. For each committee, we use complementary counting. There are $\binom{15}{4} = 1365$ ways to choose 4 students for the committee, $\binom{5}{4} = 5$ ways of choosing a committee of only males, and $\binom{10}{4} = 210$ ways of choosing a committee of only females. Therefore, there are $1365 - 5 - 210 = 1150$ ways to choose each committee and so $1150^2 = 1322500$ to choose both committees.

Answer: 1322500

Problem 15 Solution

First note that the fourth polygon must have an interior angle of $360 - 90 - 90 - 60 = 120$ and hence must be a hexagon. Therefore the perimeter of the entire shape is the sum of the perimeter of each of the four polygons minus the 4 shared edges. Note each shared edge is counted twice. Hence the perimeter is $6 + 4 + 4 + 3 - 4 \cdot 2 = 9$.

Answer: 9

Problem 16 Solution

Let's pretend for a second that we are only about choosing 4 teams (say A, B, C, D). This can be done in

$$\binom{12}{3}\binom{9}{3}\binom{6}{3}\binom{3}{3} = 369600$$

different ways. However, if A plays B and C plays D, we do not care about the order of teams A, B or C, D (2! ways for each), so there are in total

$$\frac{369600}{2!2!} = 92400$$

different ways this can happen.

Answer: 92400

Problem 17 Solution

Fermat's Little Theorem says that $a^6 \equiv 1 \pmod 7$ if $7 \nmid a$. So,

$$222^{555} \equiv 5^{555} \equiv 5^{6 \times 92 + 3} \equiv 5^3 \equiv 6 \pmod 7,$$

and

$$555^{222} \equiv 2^{222} \equiv 2^{6 \times 37} \equiv 1 \pmod 7,$$

so the sum is $0 \pmod 7$.

Answer: 0

Problem 18 Solution

Suppose the altitudes (from A, B, C respectively) are h_A, h_B, h_C, with $h_A : h_B : h_C = 2 : 2 : 3$. If a, b, c are the opposite sides, we then have (after canceling $1/2$) $a \cdot h_A = b \cdot h_B = c \cdot h_C$. We therefore have that the side lengths are in ratio

$$a : b : c = \frac{1}{h_A} : \frac{1}{h_B} : \frac{1}{h_C} = \frac{1}{2} : \frac{1}{2} : \frac{1}{3} = 3 : 3 : 2.$$

If the perimeter of the triangle is 24, this means the sides of the triangles are 9, 9, and 6. Using Heron's formula, the area is

$$\sqrt{12 \cdot (12 - 9)^2 \cdot (12 - 6)} = \sqrt{648},$$

so $M = 648$.

Answer: 648

Problem 19 Solution

Let $r \le s$ be the two roots. We know $rs = q$ and $r + s = -p$ by Viete's theorem. Hence

$$198 = p + q = rs - r - s.$$

Completing the rectangle,

$$(r-1)(s-1) = 199.$$

Note 199 is prime, so we must have $r-1 = 1$ and $s-1 = 199$ so $r = 2$ and $s = 200$. Hence $p = -(2+200) = -202$.

Answer: -202

Problem 20 Solution

Note that in the figure above, there are two cones sharing the same circular base. Let r be the radius of the cones and let h be the height of the smaller cone. Therefore, h and r satisfy $h^2 + r^2 = 10^2$ and $(h+9)^2 + r^2 = 17^2$. $(6, 8, 10)$ and $(8, 15, 17)$ are Pythagorean triples, so $h = 6$ and $r = 8$. Therefore, the surface area of the new figure is

$$\pi(8 \times 17) + \pi(8 \times 10) = 216\pi.$$

Thus $K = 216$.

Answer: 216

2.6 ZIML March 2017 Junior Varsity

Below are the solutions from the Junior Varsity ZIML Competition held in March 2017.
The problems from the contest are available on p.43.

Problem 1 Solution
$f(x) = -36(x^2 - 2x + 1) + 97 = -36(x-1)^2 + 97$. $f(x)$ attains the maximum value when $x = 1$, which is 97.

Answer: 97

Problem 2 Solution
A pair of circles can intersect in at most 2 points. It is possible to draw 4 circles so that each pair intersects twice. In this case there are

$$\binom{4}{2} \times 2 = 6 \times 2 = 12$$

points that lie on at least two of the circles.

Answer: 12

Problem 3 Solution
Note 2^{20} has 21 factors: $2^0, 2^1, \ldots, 2^{20}$. Further, $2^{12} = 4096 < 5000 < 8192 = 2^{13}$. Hence, 13 of the factors are less than 5000, so $21 - 13 = 8$ are larger.

Answer: 8

Problem 4 Solution
The last two digits of powers of 21 follow a pattern: 21, 41, 61, 81, 1, 21, 41, 61, 81, 1, ... which has cycle length 5. Since $2017 \equiv 2 \pmod 5$ the last two digits are 41.

Answer: 41

Problem 5 Solution

Let x_1 and x_2 be the roots of the equation. Applying Viete's Theorem to the quadratic equation $x^2 + 2kx - 3 = 0$ yields $x_1 + x_2 = -2k$ and $x_1 x_2 = -3$. Note that $4k^2 = (-2k)^2 = (x_1 + x_2)^2 = x_1^2 + 2x_1x_2 + x_2^2 = 10 + 2(-3) = 4$. This implies $k = \pm 1$. Since $k > 0$, $k = 1$.

Answer: 1

Problem 6 Solution

We use stars and bars. Start with 10 stars and $3 - 1 = 2$ bars. Set aside $3 \cdot 2 = 6$ stars to ensure $a, b, c \geq 2$. We are left with 4 stars and 2 bars to rearrange without restriction:

$$\binom{4+2}{4} = \binom{4+2}{2} = \binom{6}{2} = 15.$$

Answer: 15

Problem 7 Solution

Circle ω_1 has center $A(0, -10)$ and radius 6, and circle ω_2 has center $B(0, 15)$ and radius 4.

For two separate circles, the length of the common *external* tangent line segment is greater than the length of the common *internal* tangent line segment (*Note to reader: can you prove this fact?*). Thus, the segment $\overline{PQ}$ is the common internal tangent to ω_1 and ω_2. From the problem, P is on ω_1 and Q is on ω_2. Hence $AP = 6$ and $BQ = 4$.

We now calculate the length of $\overline{AB}$. Since both A and B are on the y-axis, the length $AB = 15 + 10 = 25$.

Extend $\overline{BQ}$ to point K so that $QK = AP = 6$, then $PQKA$ is a rectangle, and $BK = 4 + 6 = 10$. The triangle ABK is a right triangle, and $AK^2 = AB^2 - BK^2 = 625 - 100 = 525$. Therefore

$PQ = AK = \sqrt{525}$.

Therefore, the shortest segment $\overline{PQ}$ that is internally tangent to both circles has length $\sqrt{525}$. Therefore the answer is $M = 525$.

Answer: 525

Problem 8 Solution
The prime factorization is $3240 = 2^3 \cdot 3^4 \cdot 5^1$, so the number of factors is $(3+1) \cdot (4+1) \cdot (1+1) = 40$.

Answer: 40

Problem 9 Solution
Let x_1 and x_2 be the two roots of the equation. Then, using Viete's Theorem we have $x_1 + x_2 = 6 - a$ and $x_1 x_2 = a$, so

$$x_1 + x_2 + x_1 x_2 = 6.$$

Complete the rectangle:

$$x_1 x_2 + x_1 + x_2 + 1 = 7,$$

so

$$(x_1 + 1)(x_2 + 1) = 7.$$

Assume $x_1 \leq x_2$. As 7 is prime we get that $x_1 + 1 = 1$ and $x_2 + 1 = 7$, or $x_1 + 1 = -7$ and $x_2 + 1 = -1$. The two roots are either 0 and 6, or -8 and -2. If the roots are 0 and 6, then $a = 0$. Hence, the roots must be -8 and -2, so $a = 16$.

Answer: 16

Problem 10 Solution
Note $2^2 - 1 = 3$, $2^3 - 1 = 7$, $2^5 - 1 = 31$, and $2^7 - 1 = 127$ are all prime numbers, but $2^{11} - 1 = 2047 = 23 \times 89$ is composite.

Answer: 11

Problem 11 Solution

First assume, 3 of the vertices are contained in the same square face. Label the cube $ABCD$-$A'B'C'D'$ so that the vertices chosen are A, B, C, and one of A', B', C', D'. Note $ABC - A' \cong ABC - C'$, so this leads to 3 different tetrahedra. The only other possibility (up to congruence) is (again using the labeling $ABCD$-$A'B'C'D'$) AC-$B'D'$.

Answer: 4

Problem 12 Solution

Firstly, we can place the red marbles in a line as shown below:

$$-R-R-R-R-R-$$

As indicated by the dashes above, this creates 6 spaces to place non-red colored marbles in the line. Of the 6 available spaces, we want to reserve 3 spaces for blue marbles. This can be done in

$$\binom{6}{3} = 20$$

ways. For any choice of arrangement of blue and red marbles, we will observe that there will be 3 remaining spaces for the remaining green marbles.

$$BRBRBR - R - R-$$

Therefore, there are only 20 such rearrangements possible.

Answer: 20

Problem 13 Solution

Note $|x^2 + 1| = x^2 + 1$ because squares are always nonnegative. If $x \geq 1$ we have $x^2 + 1 = 2x - 2$ so $x^2 - 2x + 3 = 0$. Note the discriminant is negative, so this has no real roots. If $x < 1$ we have

$x^2 + 1 = 2 - 2x$. Hence $x^2 + 2x - 1 = 0$ and we can solve for x to get $x = -1 \pm \sqrt{2}$ using the quadratic formula. Thus $A + B = 1$.

Answer: 1

Problem 14 Solution

We can assume the red piece is placed before the black piece. We break the problem into cases based on whether the red piece is in (i) a corner, (ii) an edge, or (iii) in the center squares of the checkerboard.

There are 4 corners and each corner allows the black checker piece to be in one of 3 adjacent spaces. Therefore, there are

$$4 \times 3 = 12$$

ways to arrange the pieces in this case. There are $6 \times 4 = 24$ edge locations and each allows the black checker piece to be in one of 5 adjacent spaces. Thus, there are

$$24 \times 5 = 120$$

ways to arrange the checker pieces with the red piece on an edge. There are 36 inside locations and each allows the black checker piece to be in one of 8 adjacent spaces. Hence, there are

$$36 \times 8 = 288$$

different arrangements in this case. Therefore, there are

$$12 + 120 + 288 = 420$$

ways to arrange the two checker pieces on the board so that they are adjacent to each other.

Answer: 420

Problem 15 Solution

Since A has 9 divisors, it must be of the form p^8 or $p^2 \cdot q^2$ for primes p, q. Since $\text{lcm}(A,B) = 3^2 \cdot 5 \cdot 7^2$, we must have $A = 3^2 \cdot 7^2 = 441$. Since $5 \nmid A$ it must be the case that $5 \mid B$, and since $\gcd(A,B) = 7$, $7 \mid B$. Since B has 4 factors, this implies that $B = 5 \cdot 7 = 35$. Thus $A + B = 476$.

Answer: 476

Problem 16 Solution

First note $\triangle ABQ \cong \triangle BCP$, therefore $\angle BAT = \angle TBC$. Consider $\triangle ABT$. Since $\angle BAT = \angle TBC$, $\angle BAT + \angle ABT = 90°$, and thus $\angle ATB = 90°$. This implies $\angle ATP = 90°$ as well.

Answer: 90

Problem 17 Solution

Let $y = x^2 + x + 1$, the expression becomes

$$y(y+1) - 12 = y^2 + y - 12 = (y+4)(y-3) = 0.$$

Changing back to x we get

$$(x^2 + x + 5)(x^2 + x - 2) = (x^2 + x + 5)(x - 1)(x + 2) = 0,$$

so the smallest solution is $x = -2$.

Answer: -2

Problem 18 Solution

Let $ABCD$ be the trapezoid, with $\overline{AD}$ the larger base and diagonal $AC = 10$. Further, let $\overline{BF}$ be the altitude from B and let E be a point on the extension of DA so that $BCAE$ is a parallelogram.

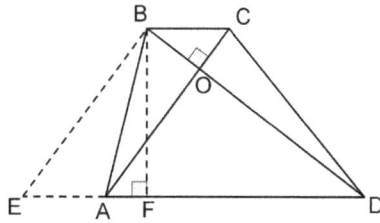

First note that $BE = AC = 10$. As $BF = 8$ we have that $EF = 6$ from the Pythagorean theorem. Next note that $\triangle BFE \sim \triangle DBE$ so

$$\frac{EF}{BE} = \frac{BE}{DE} \Rightarrow \frac{6}{10} = \frac{10}{DE} \Rightarrow DE = \frac{50}{3}.$$

Then, as $DE = AD + AE = AD + BC$ we have that the area of the trapezoid is

$$\frac{AD + BC}{2} \times BF = \frac{50}{6} \times 8 = \frac{200}{3}.$$

Thus $p + q = 203$.

Answer: 203

Problem 19 Solution

Since $6 = 2 \times 3$, we want a number that is a multiple of 2, 3, and 5. For divisibility by 2 and 5, the last digit must be 0. One of our digits is 0, and if we want the smallest number, we should try 1 as our other digit. To be divisible by 3, the sum of the digits must be divisible by 3, so we need at least 3 ones. The smallest number, with three 1's and five 0's, that ends in 0 is thus 10000110.

Answer: 10000110

Problem 20 Solution

He needs to draw a black face card in order for you to win $2. There are three face cards, (J, Q, K) and two black suits (Clubs, Spades). Thus, there are

$$2 \times 3 = 6$$

black face cards in the deck of 52 cards. Therefore, the probability of earning \$2 is:

$$\frac{6}{52} = \frac{3}{26}.$$

Hence $p + q = 29$.

Answer: 29

2.7   ZIML April 2017 Junior Varsity

Below are the solutions from the Junior Varsity ZIML Competition held in April 2017.
The problems from the contest are available on p.49.

Problem 1 Solution
The prime factorizations of 360 and 525 are $2^3 \cdot 3^2 \cdot 5$ and $3 \cdot 5^2 \cdot 7$, respectively. So, their lcm is $2^3 \cdot 3^2 \cdot 5^2 \cdot 7$. The smallest multiple of this lcm that is a perfect square is $2^4 \cdot 3^2 \cdot 5^2 \cdot 7^2 = 176400$.

Answer: 176400

Problem 2 Solution
Regular pentagons have angles of $108°$, so $\angle TAP = 108°$. Similar reasoning for equilateral triangles gives $\angle BAT = 60°$. If $\triangle TAB$ is drawn inside the pentagon we get $\angle BAP = 108° - 60° = 48°$. If $\triangle TAB$ is outside the pentagon we get $\angle BAP = 108° + 60° = 168°$. As these are the only two possibilities, the largest possible is $168°$.

Answer: 168

Problem 3 Solution
There are 5! ways to order all the boys. Doing this creates 6 spaces where the girls could finish, with at most one girl in each space as they cannot finish right after one another. Hence there are $6 \cdot 5 \cdot 4$ ways to choose spaces for the girls. Therefore the final answer is
$$5! \cdot 6 \cdot 5 \cdot 4 = 14400,$$
possible ways.

Answer: 14400

Problem 4 Solution

Let $y = 4$, so the polynomial is $y^2 + (x - x^2)y - x^3$, a quadratic in y, which factors as $(y - x^2)(y + x)$. Therefore $p(x) = (4 - x^2)(4 + x) = (2 - x)(2 + x)(4 + x)$. Thus the roots are $\pm 2, -4$, so $r = 2$ and $s = -4$, and $|r| + |s| = 2 + 4 = 6$.

Answer: 6

Problem 5 Solution

Since $\angle ABC = 80°$, we have $\overset{\frown}{CA} = 160°$. Hence

$$\overset{\frown}{AB} = \frac{6}{10} \cdot 160° = 96°.$$

A full circle is $360°$, so

$$\overset{\frown}{BC} = 360° - 160° - 96° = 104°.$$

This gives $\angle BAC = 104° \div 2 = 52°$.

Answer: 52

Problem 6 Solution

Any rectangle is made up of two horizontal lines and two vertical lines. In fact, any two horizontal lines and vertical lines from the diagram determine a unique rectangle. As there are 6 horizontal and 7 vertical lines, there are

$$\binom{6}{2} \times \binom{7}{2} = 15 \times 21 = 315$$

total rectangles.

Answer: 315

Problem 7 Solution

Let $z = x^2 + 4x + 5$. Then our equation becomes $z^2 + 5xz + 6x^2 = 0$. The left hand side can be factored:

$$z^2 + 5xz + 6x^2 = (z + 2x)(z + 3x).$$

Plugging back in x we have

$$(x^2 + 4x + 5 + 2x)(x^2 + 4x + 5 + 3x)$$
$$= (x^2 + 6x + 5)(x^2 + 7x + 5)$$
$$= (x+1)(x+5)(x^2 + 7x + 5).$$

The equation $x^2 + 7x + 5 = 0$ does not have integer solutions, hence the integer roots are -1 and -5, with product 5.

Answer: 5

Problem 8 Solution

First note the areas of $\triangle ADE$ and $\triangle BEC$ are equal (areas of $\triangle ABD$ and $\triangle ABC$ share the same height and base so have equal areas, then subtract off $\triangle ABE$). Assume their areas are each x. Then since $\overline{AB} \parallel \overline{CD}$, $\triangle ABE \sim \triangle CDE$ with ratio of sides $10 : 15$, so we know that $BE : DE = 2 : 3$. Thus, as triangles $\triangle ADE$ and $\triangle ABE$ share a height from A, the area of $\triangle ABE$ is $\dfrac{2x}{3}$. A similar argument gives $\triangle CDE$ has area $\dfrac{3x}{2}$. Hence, trapezoid $ABCD$ has area

$$x + x + \frac{2x}{3} + \frac{3x}{2} = \frac{25x}{6},$$

so $\triangle ADE$ has area $\dfrac{6}{25}$ of the full trapezoid. Hence $P + Q = 6 + 25 = 31$.

Answer: 31

Problem 9 Solution

$15 \equiv -1 \pmod{16}$ and similarly $33 \equiv 1 \pmod{16}$, hence

$$15^{999} + 33^{100} \equiv (-1)^{999} + 1^{100} \equiv -1 + 1 \equiv 0 \pmod{16},$$

so the remainder is 0.

Answer: 0

Problem 10 Solution

Use $x^3 = -3x + 9$ to reduce the exponents.

$$
\begin{aligned}
x^7 + 54x(x-1) &= x(-3x+9)^2 + 54x^2 - 54x \\
&= 9x^3 - 54x^2 + 81x + 54x^2 - 54x \\
&= 9x^3 + 27x \\
&= 9(x^3 + 3x) \\
&= 9 \times 9 \\
&= 81.
\end{aligned}
$$

Answer: 81

Problem 11 Solution

There are $\binom{8}{3} = 56$ total outcomes with 5 tails and 3 heads. Of these, to count how many have 3 heads together, we can group all 3 heads and count them as a group; thus, there are 6 outcomes with all 3 heads in a row. Hence the probability is

$$
\frac{6}{56} = \frac{3}{28},
$$

so $Q - P = 28 - 3 = 25$.

Answer: 25

Problem 12 Solution

Recall that a number is congruent to the alternating sum of its digits (working right to left) mod 11. Hence the number above is congruent to

$$
7 - 1 + 0 - 2 + 7 - 1 + \cdots + 7 - 1 + 0 - 2 \equiv 4 \cdot 2017 \equiv 4 \cdot 4
$$
$$
\equiv 16 \equiv 5 \pmod{11},
$$

so the remainder is 5.

Answer: 5

Problem 13 Solution

Let $x_1 > x_2$ be the two roots. By Viete's formulas, $x_1 + x_2 = -m$ and $x_1 x_2 = 1 - m$. Substitute the first equation into the second to get $x_1 x_2 - x_1 - x_2 = 1$. Now add 1 and factor the left-hand side to get $(x_1 - 1)(x_2 - 1) = 2$. This implies that $x_1 = 3$ and $x_2 = 2$. Hence $m = -(x_1 + x_2) = -5$.

Answer: -5

Problem 14 Solution

Suppose Bob arrives x hours after 5 pm (so $0 \le x \le 1$) and Charlie arrives y hours after 5 pm (so $0 \le y \le 2$), and consider the point (x, y) in the coordinate plane. All the possible points form a 1×2 rectangle with area 2. The region corresponding to Charlie arriving before Bob is the part of this rectangle with $y < x$, which is a triangle with base and height 1. As this triangle has area 0.5, or

$$\frac{0.5}{2} = 0.25$$

of the entire rectangle, the probability is 0.25.

Answer: 0.25

Problem 15 Solution

We know the side lengths for a Pythagorean triple containing 17. Recall all Pythagorean triples can be written in the form $(m^2 - n^2, 2mn, m^2 + n^2)$ for integers m, n. Since 17 is odd and we want the largest possible area, we try $17 = m^2 - n^2$. Factoring we have

$$17 = (m - n)(m + n) \Rightarrow m - n = 1, \ m + n = 17$$

as 17 is prime. Thus, $m = 9$ and $n = 8$, implying that the triangle has side lengths $17, 144, 145$, and thus area $\frac{1}{2} \cdot 17 \cdot 144 = 1224$.

Answer: 1224

Problem 16 Solution

Using the angle bisector theorem we have $AB : AC = BE : EC = 3 : 4$ so $AB = 9$. Similarly we find $AD = 15$. Hence the perimeter is $9 + 14 + 18 + 15 = 56$.

Answer: 56

Problem 17 Solution

Divide everything by x^2: $2\left(x^2 + \frac{1}{x^2}\right) - 9\left(x + \frac{1}{x}\right) + 14 = 0$.

Let $y = x + \frac{1}{x}$, then $y^2 = x^2 + \frac{1}{x^2} + 2$ and $2(y^2 - 2) - 9y + 14 = 0$.

Solving for y we obtain solutions $y = 2$ and $y = \frac{5}{2}$, which yield $x = 1$, $x = 2$, and $x = 1/2$, a total of 3 different roots.

Answer: 3

Problem 18 Solution

Note we only need to keep track of the units digit and figure out which digit appears consecutively first. The units digits of the Fibonacci sequence are:

$$1, 1, 2, 3, 5, 8, 3, 1, 4, 5, 9, 4, 3, 7, 0, 7, 7, \ldots,$$

so we see that Alex should choose the digit $D = 7$.

Answer: 7

Problem 19 Solution

Note the solid $EFG - JKL$ is what remains after removing the tetrahedra $J - AEG, K - BEF, L - CFG, S - JKL$ from the original tetrahedron $S - ABC$. Since these four removed tetrahedra are formed from midpoints, they are all congruent and themselves regular tetrahedra. These tetrahedra have side lengths that are half of the original tetrahedron, so they have volume that is

$$\left(\frac{1}{2}\right)^3 = \frac{1}{8}$$

of the original. Hence solid $EFG - JKL$ has volume

$$100 - 4 \cdot \frac{1}{8} \cdot 100 = 100 - 50 = 50.$$

Answer: 50

Problem 20 Solution

Let A, B, C be respectively events that the three friends get 0 books. Our final answer will be

$$3^7 - n(A \cup B \cup C),$$

as there are 3^7 total ways to give out the books, and we need to subtract of the number of ways at least one of the friends gets zero books. We have $n(A) = n(B) = n(C) = 2^7$, $n(A \cap B) = n(A \cap C) = n(B \cap C) = 1^7$ and $n(A \cap B \cap C) = 0$. Therefore using the Principle of Inclusion Exclusion,

$$n(A \cup B \cup C) = 3 \cdot 2^7 - 3.$$

Hence the total number of ways to give out the books is

$$3^7 - 3 \cdot 2^7 + 3 = 2187 - 192 + 3 = 1806.$$

Answer: 1806

2.8 ZIML May 2017 Junior Varsity

Below are the solutions from the Junior Varsity ZIML Competition held in May 2017.

The problems from the contest are available on p.55.

Problem 1 Solution

We must have $4 \mid \overline{7b}$, so $b = 2$ or 6. The alternating sum of digits is a multiple of 11, so $11 \mid a - 3 + 5 - 7 + b = a + b - 5$. Since $0 \leq a, b \leq 9$, we have $a + b - 5 = 0$ or $a + b - 5 = 11$. We see that, if $b = 2$, then $a = 3$ is a solution to the first equation. If $b = 6$, there is no a in the range $0 \leq a \leq 9$ that satisfies either equation. Hence $a = 3$ and $b = 2$, so the number is 33572.

Answer: 33572

Problem 2 Solution

Let $x = AD$. Note that $\triangle ADE$ is also equilateral. Thus the perimeter of $\triangle ADE = 3x$ and the perimeter of trapezoid $DECB$ is $x + 2(16 - x) + 16 = 48 - x$. Since these two perimeters are equal, $3x = 48 - x$ so $x = 12$. Thus the area of $\triangle ADE$ is $\dfrac{12^2\sqrt{3}}{4} = 36\sqrt{3}$ so $M = 36$.

Answer: 36

Problem 3 Solution

Consider two cases, either both rings are on the same finger or they are on different fingers. If they are on the same finger, we must choose the finger and order the rings, so there are $2! \cdot 4 = 8$ ways. If they are on different fingers there are $4 \cdot 3 = 12$ ways to choose the fingers for each ring. Hence there are

$$2! \cdot 4 + 4 \cdot 3 = 20$$

total outcomes.

Answer: 20

Problem 4 Solution

Using Viete's formulas, we know $x_1 + x_2 = \dfrac{8}{17}$ and $x_1 x_2 = \dfrac{-2}{17}$, so

$$x_1^2 + x_2^2 = (x_1 + x_2)^2 - 2x_1 x_2 = \left(\frac{8}{17}\right)^2 - 2\left(\frac{-2}{17}\right)$$
$$= \frac{64 + 4 \cdot 17}{17^2} = \frac{132}{289}.$$

Hence $P + Q = 132 + 289 = 421$.

Answer: 421

Problem 5 Solution

There are 5! ways to arrange the 5 consonants, and they create $5 + 1 = 6$ spaces that can be used for placing the vowels. Hence there are

$$5! \cdot 6 \cdot 5 \cdot 4 = 14400$$

total rearrangements.

Answer: 14400

Problem 6 Solution

Note that, for $n = 1, 2, 3, 4, 5$, we have $n^2 \equiv 1, 4, 4, 1, 0 \pmod 5$, and this sequence repeats thereafter. Hence

$$1^2 + 2^2 + \cdots + 49^2 \equiv 10 \cdot (1 + 4 + 4 + 1 + 0) \equiv 0 \pmod 5.$$

Alternatively

$$1^2 + 2^2 + \cdots + 49^2 = \frac{49 \cdot 50 \cdot 99}{6} = 49 \cdot 25 \cdot 33$$

so as 25 is a multiple of 5, the remainder is 0 when divided by 5.

Answer: 0

Problem 7 Solution

Let $[ABC]$ denote the area of $\triangle ABC$. Let O be the intersection of $\overline{AM}$ and $\overline{DE}$. Then $[AOE] = [MOD]$, since $ADME$ is a trapezoid (because $\overline{ME} \parallel \overline{AD}$). Also $[ABM] = [ACM]$ because M is the midpoint of $\overline{BC}$. Therefore,

$$[BMOE] = [ABM] - [AOE] = [ACM] - [MOD] = [AODC],$$

so

$$[BDE] = [BMOE] + [MOD] = [AODC] + [AOE] = [AEDC].$$

This gives that $[BDE] = [ABC] \div 2 = 90 \div 2 = 45$.

Answer: 45

Problem 8 Solution

First note

$$x^2 + x - 14 + \frac{1-x}{x^2} = x^2 + x - 14 - \frac{1}{x} + \frac{1}{x^2}.$$

Squaring $x - \frac{1}{x} = 5$ we get $x^2 - 2 + \frac{1}{x^2} = 25$ or $x^2 + \frac{1}{x^2} = 27$.
Thus

$$x^2 + x - 14 - \frac{1}{x} + \frac{1}{x^2} = x^2 + \frac{1}{x^2} + x - \frac{1}{x} - 14 = 27 + 5 - 14 = 18.$$

Hence $L = 18$.

Answer: 18

Problem 9 Solution

In total, there are $9 \cdot 10^4 = 90000$ five-digit numbers as we choose any digit except for 0 as the first digit. Similarly there are $8 \cdot 9^4 =$

52488 five-digit numbers that do not contain the digit 5. Hence in total there are

$$9 \cdot 10^4 - 8 \cdot 9^4 = 90000 - 52488 = 37512$$

five-digit numbers containing the digit 5 at least once.

Answer: 37512

Problem 10 Solution
To have $10 = (9+1) = (4+1)(1+1)$ factors, we see a number must be of the form p^9 or $p^4 q$ for primes p, q. The smallest such number is $48 = 2^4 \times 3$.

Answer: 48

Problem 11 Solution
Using symmetry we look at spheres with centers across the diagonal:

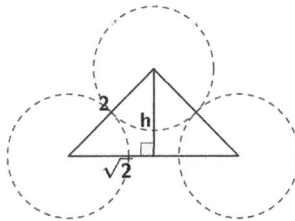

Let $h+1$ denote the height of the center of the fifth sphere off the ground. Setting up right triangles gives us $2^2 = 2 + h^2$, so solving for h gives $\sqrt{2}$. Hence the height is $\sqrt{2} \approx 1.414 \approx 1.4$.

Answer: 1.4

Problem 12 Solution
We know $\sqrt{x+3} = \sqrt{3x+2} - 1$, so by squaring, $x+3 = 3x - 2 - 2\sqrt{3x-2} + 1$. Thus $x - 2 = \sqrt{3x-2}$, and square again to get

$x^2 - 7x + 6 = 0$. Factoring we get $x = 1$ or $x = 6$. $x = 1$ is an extraneous solution. Therefore $x = 6$ is the only solution, thus the sum of the solutions is 6.

Answer: 6

Problem 13 Solution

Since we do not care about the actual people getting of at each floor, we consider them all to be identical. Hence we are dividing 10 indistinguishable objects (the people) into 5 distinguishable groups (the floors). This is a standard stars and bars (or sticks and stones) problem, with 10 stars and $5 - 1 = 4$ bars. Hence there are

$$\binom{10 + 5 - 1}{10} = \binom{14}{10} = \binom{14}{4} = 1001$$

total possible outcomes.

Answer: 1001

Problem 14 Solution

Rotate $\triangle PDC$ $90°$ about C. Call the point P is rotated to P'. Note that $\triangle PQC$ is congruent to $\triangle P'QC$. Hence $\angle P'CQ = \angle PCQ$. Since $PCP' = 90°$, $\angle PCQ = 90 \div 2 = 45°$.

Answer: 45

Problem 15 Solution

Note that

$$\frac{1}{ab} + \frac{1}{ac} + \frac{1}{bc} = \frac{a + b + c}{abc}.$$

Using Viete's, we have $a + b + c = -20/2 = -10$ and $abc = -50/2 = -25$. Hence

$$\frac{1}{ab} + \frac{1}{ac} + \frac{1}{bc} = \frac{a + b + c}{abc} = \frac{-10}{-25} = \frac{2}{5} = 0.4.$$

Answer: 0.4

Problem 16 Solution

There are 25 primes under 100. Divide the set $\{1, 2, \ldots, 100\}$ into 26 categories: the number 1 itself is a category, and each of the other categories contain the multiples of one of the primes. It is noted that these categories overlap, but together they cover all the set $\{1, 2, \ldots, 100\}$. If we choose 27 numbers, it is guaranteed that there must be two of the numbers that belong to the same category, so these two numbers are not relatively prime. So 27 is sufficient. However, if we only select 26 numbers, the worst case scenario is that we have chosen all the prime numbers plus the number 1, and all those are pairwise relatively prime. Therefore 27 is the minimum value.

Answer: 27

Problem 17 Solution

First note we can think of the balls as 5 green and $6 + 4 = 10$ not green.

Our denominator, the total number of outcomes, is either 2 or 3 balls green. Since we are picking without replacement, this is

$$\binom{5}{2}\binom{10}{1} + \binom{5}{3}\binom{10}{0} = 100 + 10 = 110.$$

Similarly, our numerator is the number of ways to get exactly 3 green balls:

$$\binom{5}{3}\binom{10}{0} = 10.$$

Hence the probability is

$$\frac{10}{110} = \frac{1}{11}$$

so $Q - P = 11 - 1 = 10$.

Answer: 10

Problem 18 Solution

Multiply by $5xy$ to get $5x + 5y = xy$ or $xy - 5x - 5y = 0$. We can add 25 to complete the rectangle and get $(x-5)(y-5) = 25$. The solutions depend on how you factor 25. We have two solutions with $0 < x < y$: $x - 5 = 1$ and $y - 5 = 25$, or $x - 5 = 5$ and $y - 5 = 5$. These solutions which give $x = 6$, $y = 30$, or $x = y = 10$. The ratio $\frac{y}{x}$ is largest for $x = 6$, and $y = 30$, giving $\frac{30}{6} = 5$ as the answer.

Answer: 5

Problem 19 Solution

From the given lengths it is easy to see that $\triangle ABP$ is a right triangle, and $\angle ABP = 90°$. So AD is the diameter. By Power of a Point, $AP \cdot PC = BP \cdot PD$, so $PD = 10 \times 5.4 \div 6 = 9$. In right triangle ABD, apply the Pythagorean Theorem to get $AD = \sqrt{AB^2 + AD^2} = \sqrt{8^2 + 45^2} = 17$. Hence the radius is $17 \div 2 = 8.5$.

Answer: 8.5

Problem 20 Solution

Substitute $y = mx + \sqrt{2}$ into $x^2 + y^2 - 1 = 0$ to get $(m^2 + 1)x^2 + 2\sqrt{2}mx + 1 = 0$. This has discriminant $\Delta = 4(m^2 - 1)$. $\Delta \geq 0$ when $-1 \leq m \leq 1$, so $b - a = 1 - (-1) = 2$.

Answer: 2

2.9 ZIML June 2017 Junior Varsity

Below are the solutions from the Junior Varsity ZIML Competition held in June 2017.

The problems from the contest are available on p.61.

Problem 1 Solution

There are 9 one-digit numbers, $9 \times 10 = 90$ two-digit numbers, $9 \times 10^2 = 900$ three-digit numbers, and 1 four-digit number, so in total you write

$$1 \cdot 9 + 2 \cdot 90 + 3 \cdot 900 + 4 \cdot 1 = 2893$$

digits.

Answer: 2893

Problem 2 Solution

Squaring both sides we get

$$\sqrt{5-x} \cdot \sqrt{2+x} = \sqrt{10} \Rightarrow (5-x)(2+x) - 10.$$

By either expanding and solving or guessing we see that $x = 0$ and $x = 3$ are solutions (and we know a quadratic equation has at most 2 solutions.) Double checking, both of these solutions work, so there are 2 real solutions.

Answer: 2

Problem 3 Solution

If we add 3 to this number, the result will be divisible by 4, 5, and 6. The least common multiple of 4, 5, and 6 is 60, so the number we want is $60 - 3 = 57$.

Answer: 57

Problem 4 Solution

Extend the sides $\overline{BC}$, $\overline{DE}$, $\overline{FG}$ and $\overline{HA}$ to both directions to make a big square. The added corners of this square are isosceles right triangles with hypotenuses $8\sqrt{2}$, so their legs are all 8.

Let $x = BC$, so the big square's side length is $x + 16$. The added corners have a total area of two squares of side 8, so the added area is 128. Thus

$$(x+16)^2 = 356 + 128 = 484$$

so $x + 16 = 22$ and $x = 6$.

Answer: 6

Problem 5 Solution

Any number and the sum of its digits are congruent modulo 9. Thus their difference must be a multiple of 9. Hence $2 + 0 + d + 0 = d + 2$ must be a multiple of 9, so $d = 7$. One such number is 2080: $2080 - (2 + 0 + 8 + 0) = 2070$.

Answer: 7

Problem 6 Solution

$396 = 4 \times 9 \times 11$, so we check divisibility by 4, 9, and 11. Divisibility by 4 means the last two digits must be divisible by 4, so z is 0, 4, or 8. Using the divisibility rules for 9 and 11, these cases lead (respectively) to the six-digit numbers: 453420, 413424, and 373428, with the smallest being 373428.

Answer: 373428

Problem 7 Solution

Note the shaded area is equal to the sum of the areas of the two small semicircles and $\triangle ABC$, minus the area of the large semicircle. The sum of the areas of the two smaller semicircles is

$$\frac{1}{2}\left(\frac{\pi}{4}(AB^2+BC^2)\right)=\frac{\pi}{8}(AB^2+BC^2)=\frac{\pi}{8}AC^2,$$

which is exactly the area of the large semicircle. Therefore, the shaded area is equal to the area of $\triangle ABC$, which is

$$\frac{1}{2}\cdot 5\cdot 12 = 30,$$

so the answer is 30.

Answer: 30

Problem 8 Solution
Either $x^2-3x+1 = x^2+x+2$ or $x^2-3x+1 = -x^2-x-2$. In the first case, $4x = -1$ so $x = -1/4$ (double checking, this satisfies the original equation). In the second case we have $2x^2-2x+3 = 0$, which has discriminant $\Delta = -20$ so has no real solutions. Hence $x = -1/4 = -0.25$ is the only solution.

Answer: -0.25

Problem 9 Solution
Grouping the German, Spanish, and French books, there are $3! = 6$ ways to arrange them as groups. Then there are $2!$, $3!$, and $4!$ ways, respectively, to order the German, Spanish, and French books themselves. This gives a total of

$$3! \times (2! \times 3! \times 4!) = 6 \times (2 \times 6 \times 24) = 1728$$

different arrangements.

Answer: 1728

Problem 10 Solution
Let $[XYZ]$ represent the area of $\triangle XYZ$. Since $[CBD] = [BAE] =$

$[ACF]$, we get

$$\frac{[CBD]}{[ABC]} = \frac{[BAE]}{[ABC]} = \frac{[ACF]}{[ABC]},$$

thus (using the fact that the triangles in each fraction share the same height)

$$\frac{125}{CA} = \frac{120}{BC} = \frac{35}{AB}.$$

Simplifying,

$$\frac{25}{CA} = \frac{24}{BC} = \frac{7}{AB}.$$

Since AB, BC, CA are all integers, the minimum possible values are $AB = 7$, $BC = 24$, and $CA = 25$. Hence $\triangle ABC$ is a right triangle, with area $\frac{1}{2} \times 7 \times 24 = 84$.

Answer: 84

Problem 11 Solution

First note that if $a = -1$ the equation becomes $-2x - 2 = 0$, so is a linear equation with one solution.

If $a \neq -1$, the equation is quadratic, so to have only one real solution the discriminant Δ must be 0. Hence

$$\Delta = (a-1)^2 - 4(a+1)(-2) = a^2 + 6a + 9 = (a+3)^2 = 0,$$

and $a = -3$.

Therefore the final answer is $-1 - 3 = -4$.

Answer: -4

Problem 12 Solution

Consider the circle divided into eighths as in the diagram below.

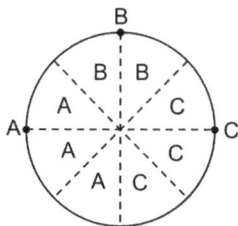

Note each eighth is labeled with which point is closest to if E is chosen inside that eighth. Hence the probability is

$$\frac{3}{8} = 0.375 = 37.5\%,$$

so $P = 37.5$.

Answer: 37.5

Problem 13 Solution
Let $z = x^2 + 3x + 1$ so the equation becomes

$$z(z+3) + 2 = z^2 + 3z + 2 = (z+1)(z+2) = 0.$$

Thus $z = -1$ or $z = -2$.

If $z = -1$, $x^2 + 3x + 1 = -1$, so $(x+1)(x+2) = 0$ and $x = -1$ or $x = -2$.

If $z = -2$, $x^2 + 3x + 1 = -2$, so $x^2 + 3x + 3 = 0$. This equation has no real roots.

Hence, the only two solutions are -1 and -2, and their difference is 1.

Answer: 1

Problem 14 Solution
Let A be the origin, and B be the center of the circle. Note the circle has radius $2\sqrt{2}$ and $AB = 4$. Hence the tangent lines to the

circle will form angles of 45° with AB (as a right triangle with hypotenuse 4 and side $2\sqrt{2}$ must be a 45-45-90 triangle). It is easy to see that the tangent lines touch the circle at $(2,\pm2)$.

The shortest path thus starts at the origin, follows one tangent line to the circle, then loops around the circle to the other tangent line and back to the original. As we are traveling $270/360 = 3/4$ of the circle, the total distance is therefore

$$2\sqrt{2}+\frac{3}{4}\cdot 2\pi\cdot 2\sqrt{2}+2\sqrt{2}\approx 4\sqrt{2}+9\sqrt{2}=13\sqrt{2}$$

using the approximation $\pi \approx 3$. Therefore our final answer is

$$13\sqrt{2}\approx 13\times 1.414\approx 18.382\approx 18$$

as needed.

Answer: 18

Problem 15 Solution
Using Viete's Theorem, $abc+bcd+cda+dab = -2016$, and $abcd = -2015$, so

$$\frac{1}{a}+\frac{1}{b}+\frac{1}{c}+\frac{1}{d}=\frac{abc+bcd+cda+dab}{abcd}=\frac{2016}{2015}.$$

Hence $P+Q = 2016+2015 = 4031$.

Answer: 4031

Problem 16 Solution
Let F be such that $\overline{AD}$ is tangent to the circle at F. Since $AB = 14$, the radius of the circle is 7. Hence, $AM = 7$ and $DF = 24-7 = 17$. By Power of a Point, $DF^2 = DM\cdot DE$. As $DM = \sqrt{7^2+24^2} = 25$,

$$DE = \frac{17^2}{25} = \frac{289}{25} = 11.56,$$

so $DE \approx 11.6$.

Answer: 11.6

Problem 17 Solution

We consider balls of each of the three colors separately. For each color, we have 4 identical balls and 4 different boxes, so we can use the non-negative version of stars and bars (with 4 stars and $4 - 1 = 3$ bars) to get $\binom{7}{4} = 35$ ways to arrange the balls of each color. Hence in total we have $35^3 = 42875$ ways to place the balls into the boxes.

Answer: 42875

Problem 18 Solution

Since $2x = \lfloor x \rfloor^2 + 1$ must be an integer, there is an integer n such that $x = n$ or $x = n + \dfrac{1}{2}$. If $x = n$, the equation simplifies to $n^2 - 2n + 1 = (n-1)^2 = 0$ so $x = n = 1$. If $x = n + \dfrac{1}{2}$, the equation simplifies to $n^2 - 2n = 0$, which has solutions $n = 0, 2$, so $x = \dfrac{1}{2}, \dfrac{5}{2}$. Hence the sum of all solutions is $1 + \frac{1}{2} + \frac{5}{2} = 4$.

Answer: 4

Problem 19 Solution

By Fermat's Little Theorem, $2^{16} \equiv 3^{16} \equiv 5^{16} \equiv 1 \pmod{17}$. Since $1000 \equiv 8 \pmod{16}$, we have

$$2^{1000} + 3^{1000} + 5^{1000} \equiv 2^8 + 3^8 + 5^8 \pmod{17}.$$

We have $2^8 \equiv (2^4)^2 \equiv (-1)^2 \equiv 1 \pmod{17}$. Note

$$3^8 \equiv (3^4)^2 \equiv 81^2 \equiv 13^2 \equiv 16 \pmod{17},$$

and similarly

$$5^8 \equiv 625^2 \equiv 13^2 \equiv 16 \pmod{17}.$$

Adding we have that

$$2^{1000} + 3^{1000} + 5^{1000} \equiv 1 + 16 + 16 \equiv 16 \pmod{17},$$

so the remainder is 16.

Answer: 16

Problem 20 Solution

Denote the events of choosing a line by L_1, L_2, and L_3, with L_1 being the malfunctioning line, and let A be the event that your chip is defective. Then, by the law of total probability,

$$P(A) = P(A|L_1)P(L_1) + P(A|L_1^c)P(L_1^c) = \frac{1}{10} \cdot \frac{1}{3} + \frac{1}{100} \cdot \frac{2}{3} = \frac{1}{25}.$$

Hence $N + M = 1 + 25 = 26$.

Answer: 26

3. Appendix

3.1 Junior Varsity Topics Covered

Algebra

- Students should be comfortable with ratios, proportions, and their applications to problems involving work and motion, but these problems are not a main focus at this level
- Exponents and Radicals: Laws of Exponents, Simplest Radical Form for Roots
- Factoring Tricks: Sums and differences of squares, cubes, etc., Binomial and multinomial theorem, Completing the Square/Rectangle, etc.
- Solving Equations: Linear Equations, Quadratic Equations, Systems of Equations, Substitutions to rewrite higher degree equations as quadratics, Radicals, Absolute Values
- Quadratics: Graphing and Vertex Form, Maxima and Minima, Quadratic Formula, Discriminant, Vieta's Theorem for sum and product of the roots
- Polynomials: Polynomial Long Division, Remainder and Factor Theorem, Rational Root Theorem, General Vieta's Theorem

Geometry

- As a general rule students should be comfortable using algebraic techniques (linear equations, quadratic equations, systems of equations, etc.) as tools for applying the geometric concepts listed below
- Angles in Parallel Lines (corresponding angles, alternating interior/exterior angles, same-side interior/exterior angles, etc.)
- Analytic Geometry: Equations of Lines, Parabolas, and Circles, Distance Formula, Midpoint Formula, Geometric Interpretation of Slope and Angles
- Triangles: Congruence and Similarity, Pythagorean theorem, Ratios of Sides for triangles with angles of 45, 45, 90 or 30, 60, 90
- Centers in Triangles: Definitions of altitudes, medians, angle bisectors, perpendicular bisectors, Definitions and basic properties of orthocenter, centroid, incenter, circumcenter, Angle Bisector Theorem
- Interior and Exterior Angles of Polygons, including the sum of all these angles, each angle if the polygon is regular, etc.
- Areas and Perimeters of basic shapes such as triangles, rectangles, parallelograms, trapezoids, and circles, Heron's formula and formulas using inradius or circumradius for triangles
- Geometric Reasoning with Areas: Congruent shapes have the same area, Similar triangles have a ratio of areas that is the square of the ratio of their sides, Triangles with the same height have a ratio of their areas equal to the ratio of their bases, etc., Using multiple expressions of area to solve for unknowns
- Circles: Arc Length, Sector Area, Definitions for Tangent Lines and Tangent Circles, Inscribed Angles, Angles formed by intersecting chords, Power of a Point, Ptolemy's Theorem
- Solid Geometry: Surface Area and Volume for Spheres, Prisms, Pyramids, and Cones, Reasoning for more general

solids, such as combining the solids listed above or pieces of solids when cut by a plane, etc.

Counting and Probability

- Fundamental Rules: Sum and Product Rules, Permutations and Combinations
- Counting Methods: Complementary counting, Stars and bars (also called sticks and stones, balls and urns, etc.), Grouping objects that must be together, Inserting objects that must be apart into spaces between objects, etc., Principle of Inclusion and Exclusion
- Identities: Symmetry, Pascal's Identity, Hockey Stick Identity, etc. for binomial coefficients, Binomial and Multinomial Theorem, Understanding of these identities using combinatorial proofs
- Sequences: Arithmetic and Geometric Sequences and Series, Finding and understanding patterns and recursive definitions for general sequences
- Probability and Sets: Definitions for event, sample space, complement, intersection, and union, Understanding the use of Venn Diagrams
- Probability: In finite sample spaces as a ratio of the number of outcomes, In geometric sample spaces as a ratio of lengths, areas, or volumes, Axioms of Probability, Independence, Conditional Probability, Law of Total Probability

Number Theory

- Fundamental Definitions: Prime numbers, factors/divisors, multiples, least common multiple (LCM), greatest common factor/divisor (GCF or GCD), perfect squares/cubes/etc.
- Number Bases: Expressing and converting numbers in base 2, 3, 8, 16, etc, Understanding how to perform arithmetic in different bases

- Divisibility Rules for numbers such as 2, 3, 4, 5, 8, 9, 10, 11, and how to combine the rules for numbers such as 6, 22, etc.
- (Unique) Prime Factorization and how to use the prime factorization to find the number of factors, to test whether a number is a perfect square/cube/etc, to find the LCM or GCD.
- Factoring Tricks: Factors come in pairs, perfect squares have an odd number of factors, etc.
- Modular Arithmetic: Connection with remainders and applications such as "find the units digit", General rules for addition, subtraction, multiplication, and division, Extension of divisibility rules to calculating a number modulo 9, 11, etc., Fermat's Little Theorem, Euler's Totient Function and extension to Fermat's Little Theorem

3.2 Glossary of Common Math Terms

Acute Angle An angle less than $90°$.

Altitude of a Triangle A line segment connecting a vertex of a triangle to the opposite side forming a right angle. Also called the height of a triangle.

Angle A figure formed by two rays sharing a common vertex. Often measured in degrees.

Angle Bisector A line dividing an angle into two equal halves.

Arc The curve of a circle connecting two points.

Area The amount of space a region takes up. Often denoted using square brackets: area of $\triangle ABC = [ABC]$.

Arithmetic Sequence A sequence where the difference between one term and the next is constant.

Average See Mean.

Base of a Triangle One side of a triangle, often used when the altitude is drawn from the opposite side to this base.

Binomial Coefficient The symbol $\binom{n}{k} = \dfrac{n!}{k!(n-k)!}$.

Centroid of a Triangle The intersection of the three medians in a triangle.

Chord A line segment connecting two points on the outside of a circle.

Circle A round shape consisting of points that all have the same distance (called the radius) from the center of the circle.

Circumcenter of a Triangle The intersection of the three perpendicular bisectors in a triangle. Also the center of the circle that circumscribes a triangle.

Circumference The perimeter of a circle.

Circumscribe To draw a shape outside another shape so that the boundaries touch.

Coefficient The number being multiplied by a variable or power of a variable. For example, the coefficient of x^3 in $5x^5 + 4x^3 + 2x$ is 4.

Complement In probability, the complement of a set is all elements outside the set.

Composite Number A number that is not prime.

Congruent Two shapes or figures that are exactly the same.

Cube A solid figure formed by 6 congruent squares that all meet at right angles.

Deck of Cards A standard deck of cards has 52 cards. There are 4 suits (clubs, diamonds, hearts, and spades) with each suit having cards of 13 ranks (A (ace), $2, 3, \ldots, 10$, J (jack), Q (queen), and K (king)).

Degree of a Polynomial The highest power of a variable in the polynomial. For example, the degree of $2x^3 - 5x^6 + 2$ is 6.

Denominator The bottom number in a fraction.

Diagonal A line segment connecting two vertices of a shape or solid that is not an edge of the shape or solid.

Diameter A chord passing through the center of a circle. The diameter has length that is twice the radius.

Die or Dice A standard die (plural is dice) has 6 sides. Each of the 6 sides has the same chance when the die is rolled.

Digit One of $0, 1, 2, \ldots, 9$ used when writing a number.

Discriminant The expression $b^2 - 4ac$ for a quadratic equation $ax^2 + bx + c = 0$.

Distinguishable Objects Objects that are different.

Divisible A number is divisible by another number if there is no remainder when the first number is divided by the second. For example, 35 is divisible by 7.

Divisor A number that evenly divides another number. For example, 6 is a divisor of 48. Also called a factor.

Edge A line segment connecting two vertices on the outside of a shape or solid.

Equally Likely Having the same chance of occurring.

Equiangular Polygon A shape with all equal angles.

Equilateral Polygon A shape with all equal sides.

Equilateral Triangle A regular triangle, one with three equal sides and three equal angles.

Even Number A number divisible by 2.

Exponent The number another number is raised to for powers. For example, in a to the power of b (a^b), the exponent is b.

Face The shape or polygon on the outside of a solid region.

Factor of a Number A number that evenly divides another number. For example, 6 is a factor of 48. Also called a divisor.

Factorial The symbol ! where $n! = n \times (n-1) \times (n-2) \cdots \times 1$.

Fraction An expression of a quotient. For example, $\frac{1}{2}$ or $\frac{9}{7}$.

Function A function is a rule that associates exactly one output with every input. Often described using an equation.

Geometric Sequence A sequence where the ratio between one term and the next is constant.

Greatest Common Divisor/Factor (GCD/GCF) The largest number that is a divisor/factor of two or more numbers.

Incenter of a Triangle The intersection of the three angle bisectors in a triangle. Also the center of a circle that is inscribed inside a triangle.

Indistinguishable Objects Objects that are the same.

Inscribe To draw a shape inside another shape so that the boundaries touch.

Intersecting Lines or curves that cross each other.

Intersection of Two Sets The set of objects that are in both of the two sets. Denoted using $\cap$. For example, $\{2,3\} \cap \{3,4,5\} = \{3\}$.

Isosceles Triangle A triangle with two equal sides and two equal angles.

Least Common Multiple (LCM) The smallest number that is a multiple of two or more numbers.

Mean The sum of the numbers in a list divided by the how many numbers occur in the list. Also called the average.

Median The number in the middle of a list when the list is arranged in increasing order.

Median of a Triangle A line connecting a vertex in a triangle to the midpoint of the opposite side.

Midpoint The point in the middle of a line segment.

Mode The number or numbers occurring most often in a list of numbers.

Multiple A number that is an integer times another number. For example, 72 is a multiple of 8.

Numerator The top number in a fraction.

Obtuse Angle An angle between $90°$ and $180°$.

Odd Number A number not divisible by 2.

Orthocenter of a Triangle The intersection of the three altitudes in a triangle.

Parallel Lines Lines that do not intersect.

Perfect Cube A number that is another number cubed. For example, $64 = 4^3$ is a perfect cube.

Perfect Square A number that is another number squared. For example, $64 = 8^2$ is a perfect square.

Perimeter The length/distance around the outside of a shape.

Perpendicular Bisector A line perpendicular to and passing through the midpoint of a line segment.

Pi (π) A number used often in geometry. $\pi = 3.1415926\ldots \approx$ $3.14 \approx \dfrac{22}{7}$.

Polygon A shape formed by connected line segments.

Polynomial A function that is made of adding multiples of powers of a variable. For example, $f(x) = x^4 + 3x^2 + 2x - 3$.

Prime Factorization The expression of a number as the product of all its prime factors. For example, 24 has prime factorization $2 \times 2 \times 2 \times 3 = 2^3 \times 3$.

Prime Number A number whose only factors are one and itself.

Proportional Ratios Ratios that have equal values when expressed in fraction form. For example, $2 : 3$ is proportional to $8 : 12$.

Quadratic A polynomial with degree 2. Often written in the form $ax^2 + bx + c$.

Quadrilateral A shape with four sides.

Quotient The integer quantity when dividing one number by another. For example, the quotient of $38 \div 5$ is 7 as $38 = 7 \times 5 + 3$.

Radius of a Circle The distance from the center of the circle to any point on the outside of the circle.

Randomly Chosen for a group of objects. Unless specified, the chance of choosing each object is the same as any other object.

Rank of a Card See Deck of Cards.

Ratio A relation depicting the relation between two quantities. For example $2 : 3$ or $\frac{2}{3}$ denotes that for every 3 of the second quantity there are 2 of the first quantity.

Rational Number A number that can be written as a fraction.

Reciprocal One divided by the number. For example, the reciprocal of 7 is $\frac{1}{7}$.

Rectangle A quadrilateral with four right angles (an equiangular quadrilateral).

Regular Polygon A polygon with all equal sides and all equal angles (equilateral and equiangular).

Remainder The quantity left over when one integer is divided by another. For example, the remainder of $38 \div 5$ is 3 as $38 = 7 \times 5 + 3$.

Rhombus A quadrilateral with four equal sides (an equilateral quadrilateral).

Right Angle A $90°$ angle.

Right Triangle A triangle containing a right angle.

Root of a Function A value of x such that the function evaluates to zero. For example, $x = 2$ is a root of the function $f(x) = x^2 - 4$.

Sample Space In probability, the sample space is the set of all outcomes for an experiment.

Scalene Triangle A triangle with three unequal sides and three unequal angles.

Sector The region formed by an arc and the two radii connecting the ends of the arc to the center of the circle.

Sequence An ordered list of numbers.

Set An unordered collection or group of objects without repeated elements. Denoted using curly brackets. For example, $\{1,2,3,4\}$ is the set containing the integers $1,\ldots,4$.

Similar Shapes or solids that have the same angles and sides that share a common ratio.

Simplest Radical Form An expression containing a radical such that the number inside the radical is an integer that has no perfect squares.

Sphere A round solid consisting of points that all have the same distance (called the radius) from the center of the sphere.

Square A shape with four equal sides and four equal angles (a regular quadrilateral).

Subset A set of objects that is contained inside a larger set of objects. Denoted using $\subseteq$. For example $\{2,3\} \subseteq \{1,2,3,4\}$.

Suit of a Card See Deck of Cards.

Surface Area The total area of all the faces of a solid.

Tangent Line A line touching a shape or curve at exactly one point.

Trapezoid A quadrilateral with one pair of parallel sides.

Triangle A shape with three sides.

Union of Two Sets The set of objects that are in one or both of the two sets. Denoted using $\cup$. For example, $\{2,3\} \cup \{3,4,5\} = \{2,3,4,5\}$.

Venn Diagram A diagram with circles used to understand the relationship between overlapping sets.

Vertex The intersection of line segments, especially the intersection of sides or edges in a shape or solid.

Volume The amount of space a solid region takes up.

With Replacement When choosing objects with replacement, a chosen object is returned to the others allowing it to be chosen more than once.

3.3 ZIML Answers

ZIML October 2016 Junior Varsity

Problem 1:   98730   Problem 11:   4

Problem 2:   6720   Problem 12:   10

Problem 3:   32   Problem 13:   4.5

Problem 4:   9   Problem 14:   27

Problem 5:   36   Problem 15:   3

Problem 6:   −3   Problem 16:   20

Problem 7:   120   Problem 17:   450

Problem 8:   27   Problem 18:   40

Problem 9:   5   Problem 19:   −995

Problem 10:   5   Problem 20:   25

ZIML November 2016 Junior Varsity

Problem 1: 6

Problem 2: 38894

Problem 3: 4

Problem 4: 101

Problem 5: 6

Problem 6: 17

Problem 7: 62.5

Problem 8: 987652413

Problem 9: 12

Problem 10: 3

Problem 11: 4

Problem 12: -23

Problem 13: 196

Problem 14: 249

Problem 15: 241

Problem 16: 51

Problem 17: 67

Problem 18: 8

Problem 19: 16

Problem 20: 9

ZIML December 2016 Junior Varsity

Problem 1:   1944   Problem 11:   36

Problem 2:   −3   Problem 12:   5

Problem 3:   10   Problem 13:   10

Problem 4:   50004   Problem 14:   72

Problem 5:   15   Problem 15:   72

Problem 6:   9   Problem 16:   4

Problem 7:   2   Problem 17:   45

Problem 8:   40   Problem 18:   6

Problem 9:   2   Problem 19:   9

Problem 10:   621   Problem 20:   909

ZIML January 2017 Junior Varsity

Problem 1:	3	**Problem 11:**	512
Problem 2:	198000	**Problem 12:**	-1
Problem 3:	1568	**Problem 13:**	20
Problem 4:	29	**Problem 14:**	0.5
Problem 5:	11	**Problem 15:**	5
Problem 6:	16	**Problem 16:**	8
Problem 7:	5	**Problem 17:**	69
Problem 8:	14400	**Problem 18:**	26
Problem 9:	3	**Problem 19:**	90
Problem 10:	2	**Problem 20:**	4

ZIML February 2017 Junior Varsity

Problem 1:	3	Problem 11:	14
Problem 2:	4	Problem 12:	-18
Problem 3:	0	Problem 13:	48
Problem 4:	120	Problem 14:	1322500
Problem 5:	1	Problem 15:	9
Problem 6:	30129	Problem 16:	92400
Problem 7:	853776	Problem 17:	0
Problem 8:	84	Problem 18:	648
Problem 9:	22	Problem 19:	-202
Problem 10:	16	Problem 20:	216

ZIML March 2017 Junior Varsity

Problem 1:	97	Problem 11:	4
Problem 2:	12	Problem 12:	20
Problem 3:	8	Problem 13:	1
Problem 4:	41	Problem 14:	420
Problem 5:	1	Problem 15:	476
Problem 6:	15	Problem 16:	90
Problem 7:	525	Problem 17:	-2
Problem 8:	40	Problem 18:	203
Problem 9:	16	Problem 19:	10000110
Problem 10:	11	Problem 20:	29

ZIML April 2017 Junior Varsity

Problem 1:	176400	Problem 11:	25
Problem 2:	168	Problem 12:	5
Problem 3:	14400	Problem 13:	−5
Problem 4:	6	Problem 14:	0.25
Problem 5:	52	Problem 15:	1224
Problem 6:	315	Problem 16:	56
Problem 7:	5	Problem 17:	3
Problem 8:	31	Problem 18:	7
Problem 9:	0	Problem 19:	50
Problem 10:	81	Problem 20:	1806

ZIML May 2017 Junior Varsity

Problem 1:	33572	**Problem 11:**	1.4
Problem 2:	36	**Problem 12:**	6
Problem 3:	20	**Problem 13:**	1001
Problem 4:	421	**Problem 14:**	45
Problem 5:	14400	**Problem 15:**	0.4
Problem 6:	0	**Problem 16:**	27
Problem 7:	45	**Problem 17:**	10
Problem 8:	18	**Problem 18:**	5
Problem 9:	37512	**Problem 19:**	8.5
Problem 10:	48	**Problem 20:**	2

ZIML June 2017 Junior Varsity

Problem 1:	2893	**Problem 11:**	-4
Problem 2:	2	**Problem 12:**	37.5
Problem 3:	57	**Problem 13:**	1
Problem 4:	6	**Problem 14:**	18
Problem 5:	7	**Problem 15:**	4031
Problem 6:	373428	**Problem 16:**	11.6
Problem 7:	30	**Problem 17:**	42875
Problem 8:	-0.25	**Problem 18:**	4
Problem 9:	1728	**Problem 19:**	16
Problem 10:	84	**Problem 20:**	26

www.ingramcontent.com/pod-product-compliance
Lightning Source LLC
Chambersburg PA
CBHW071852200326
41519CB00016B/4342